苏州园林
Suzhou Gardens

苏州拙园为晚清最大园林。园主顾文彬（字子山）官搜藏、甚富，藏书与为画父，名过云楼，修葺颇精，楼旁老厅储置校，叠石错落有致，先生尝亲为造园，运石与修竟时正砚任浙江宁绍台道台，园之碑尝出其子顾承手，砚研查书时延宗为吴郡王云（石湖）唐而泉，顾沄（吾庐）袁云人程翊跨唐寿手设计之事，供运时诸堆石，构一亭必机稿就商乃父，其经历时扎者右供勇怀砚之砚处。

殿试后
苏州寻碍事之绍园（徐蔚吉图） 名轼情四人
为戈裕良垒。中园曹放作调查，园僻城郊，埠下者
大顿山立，右有一洞老成推转，山下有池，水洞临流，生其
大署处，是吾为戈氏所作猪徕证耳。

清朝官吏在客厅中接见属员与宾客时，侍者献茶，

陈从周文稿手迹
A page of handwritten manuscript by Chen Congzhou

（汉英对照）

苏州园林

Suzhou Gardens

陈从周 著
Chen Congzhou

策划 陈胜吾 陈馨
Planned by
Chen Shengwu & Chen Xin

翻译 朱文俊 陈威
Translated by
Zhu Wenjun & Chen Wei

上海人民出版社

目 录

6　《苏州园林》再版序（中文）

8　《苏州园林》再版序（英文）

10　代序：我的第一本书《苏州园林》（中文）

12　代序：我的第一本书《苏州园林》（英文）

15　苏州园林初步分析（中文）

35　苏州园林初步分析（英文）

81　苏州园林摄影图录（中英文）

273　苏州拙政园测绘图录（中英文）

323　苏州留园测绘图录（中英文）

374　测绘图录编后附记（中英文）

375　后记（中文）

379　后记（英文）

Table of Contents

6	Preface to the Second Edition of *Suzhou Gardens* (in Chinese)
8	Preface to the Second Edition of *Suzhou Gardens* (in English)
10	In Lieu of Preface, My First Book *Suzhou Gardens* (in Chinese)
12	In Lieu of Preface, My First Book *Suzhou Gardens* (in English)
15	Analysis of Suzhou Gardens (in Chinese)
35	Analysis of Suzhou Gardens (in English)
81	Photographs of Suzhou Gardens (in Chinese and English)
273	Architectural Survey of Zhuo Zheng Yuan (The Humble Administrator's Garden) in Suzhou (in Chinese and English)
323	Architectural Survey of Liu Yuan (The Lingering Garden) in Suzhou (in Chinese and English)
374	Notes
375	Afterwords (in Chinese)
379	Afterwords (in English)

《苏州园林》再版序

我们的老师陈从周先生之代表作《苏州园林》，1957年时由同济大学教材科印制。30年前，为其学生首得其书，至今乃景情如初：红漆布硬面，色泽沉稳；烫印书名，字体古雅；大本版面，纸质敦实；多年之尘封，却仍墨香。那是"文革"刚过，叫后辈们爱不释手。最为难忘，书中有多幅的苏州园林黑白照片，层次分明，构图优雅，有水墨画的韵味；照片题图居然是精心选择的宋词词句；一行"庭院深深深几许"，胜却无数解说和注释，动人心怀。于是学生谨存，此是可唯美，可如诗如画的书。

听先生讲的课，正是用诗情画意诠释中国园林。先生乃诗人，留下《山湖处处》一集；先生乃画家，"一寸丝一寸柔情"写照心胸情怀；先生更是独一无二的园林大家。先生从来不把中国园林看作单纯的工程技术，乃重在景物之文化内涵，重在人与景之对话，重在人对景之体验与再创造。在他眼里，中国园林之所以珍贵，是因为她和中国的诗词、绘画、戏曲融会贯通、互相映衬。中国园林正是中国文化的一种至高境界。

继《苏州园林》之后，先生有《扬州园林》、《绍兴石桥》、《园林谈丛》、《说园》等众多园林专著。每一本书，都是饱蘸笔墨，对园林艺术的精妙解析，满含着先生对中国文化的深情厚意。先生把"明轩"建到纽约大都会博物馆，先生重修了上海豫园内园，先生指点大江南北多少古典园林的修复……细细品来，其学术思想，正是开始于《苏州园林》一书，而其后之《说园》达到巅峰。

从《苏州园林》首版问世至今，五十年过去了；先生园林思想令几代学生受益，让中外学者感悟。如今我们躬逢盛世，得以重新再版此书，并以中英文双语印行，既是诗情画意的发扬光大，也是对先生最好之怀念。

吴志强

同济大学建筑与城市规划学院院长

2008年11月，陈从周先生90诞辰之际

Preface to the Second Edition of *Suzhou Gardens*

The book *Suzhou Gardens*, a masterpiece of Professor Chen Congzhou, our teacher, was printed originally by the Course Book Press of Tongji University in 1957. It is three decades ago when I first read the book. Up till now I still cannot forget the impression it made on me then: a hardcover of deep-red varnished cloth; a gilded title in graceful Chinese calligraphy; a sixteenmo print on thick art paper; and fresh ink perfume suffusing the pages, though having been shelved for years. You can imagine how much we students loved that book in those years immediately after the "Cultural Revolution". What is most unforgettable is those black-and-white pictures of Suzhou gardens, ingeniously organized and elegantly conceived with a flavor of wash painting. What is equally impressive is the meticulously-selected Song ci sentences as inscriptions for those pictures. Such poetic sentences as "Deep, deep and how much deeper is the courtyard further along?" outshine long-winded explanations or wordy notes and infuse the pictures with life and passion. I treasure this book, because it is an aesthetic book, a book brimming with deep-rooted sentiments of a poet and a painter.

Professor Chen's lectures are just intended to illustrate Chinese gardens from an approach of verse and painting. He is a poet, leaving behind him a collection of poems "Omnipresent Hills and Lakes". He is a painter, offering to the world the well-known painting "A Sprig of Willow, A Length of Tenderness", a portrayal of his artistic fervor. Above all he stands alone as a grand master of Chinese garden architect. He never regards Chinese gardens simply as the fruits of engineering technology. He highlights, however, the cultural connotation of landscapes, the dialogue between men and landscapes, and men's encounter with and recreation of landscapes. In his eye, the preciousness of Chinese gardens lies in their communication, contrast and harmony with Chinese poetry, painting and drama. In a nutshell, Chinese gardens represent the acme of Chinese culture.

Since *Suzhou Gardens* came to light, Professor Chen had many monographs on gardens published such as *Yangzhou Gardens*, *The Stone Bridges of Shaoxing*, *Talks on Gardens*, *On Gardens* and so on. Dedicated to high academic standard, each and every book goes to great length to present his incisive analysis of the art of gardens and to display his deep affection for Chinese culture. It is also worthwhile to enumerate some of his architectural achievements such as designing and building the Ming Room at the New York Metropolitan Museum, renovating the inner court of Yu Yuan Garden in Shanghai, directing the repair of many classic gardens in the north and south of China, and other projects. After a careful study of Professor Chen's career, we have found out that his academic ideas started with *Suzhou Gardens* and reached the pinnacle with *On Gardens*.

Fifty years have elapsed since the publication of *Suzhou Gardens*. His ideas on garden art have benefited several generations of students and sharpened the awareness of scholars both at home and abroad. The publication of the new bilingual (Chinese and English) edition at the present time of peace and prosperity not only enhances his distinctive style of harmonizing the art of gardens with poetry and painting, but also comes as the best token of our memory of our respected Professor Chen Congzhou.

Written on the occasion of the 90th birthday of Professor Chen Congzhou.

Wu Zhiqiang
Chancellor of the College of Architecture and City Planning
of Tongji University, Shanghai
November of 2008

代序：我的第一本书《苏州园林》

陈从周

我的第一本书，本应指我最早写作的。然而像我这种兴趣多方面的人，最初写的书并不是我的本行，例如诗人《徐志摩年谱》，完全是一次感情的冲动。还有一些零星的建筑书籍，也不过仅是偶然资料的收集。如果正式写书的话，那应该算《苏州园林》了，这是1956年完成的。也是解放后研究讨论苏州园林所出版的第一部书。

五十年代初，我在上海同济大学建筑系任教，同时又在苏州苏南工专兼课。我苏州的课是在星期六的上午。我星期五晚车去苏州，住在观前附近旅馆中，第二天清晨去沧浪亭该校上课。午梦初回，我信步园林，以笔记本、照相机、尺纸自随。真可说："兴移无洒扫，随意望莓苔"。自游，自品，俯拾得之。次日煦阳初照，叩门入园。直至午阴嘉树清园，香茗佐点，小酌山间，那时游人稀少，任我盘桓，忘午倦之侵人也。待到夕阳红半，尽一日之兴，我也上火车站，载兴而归。儿辈倚门相待，以苏州茶食迎得一笑。如今他们的年龄，正与我当年相仿佛，《苏州园林》前年在日本再版了，都已经是第二代了。

我这样每周乐此不疲，经过几年的资料累积，与所见所想，开始写我的文章。我的这些立论，并不是凭空而来，是实中求虚，自信尚有所据者。情以游兴，本来中国园林就是"文人园"，它是以诗情画意作主导思想的。因此在图片中，很自然地流露出过去所说的前人词句，我于是在每张图片下，撷了宋词题上。我将一本造园的科技书，以文学化出之。似乎是感到清新的。书出版后，受到了读者的赞誉好评，但1958年却因此受到了批判，说我士大夫的意识浓厚，我只好低头认罪，承认思想没有改造好。可是事隔近卅年，在文理相通的新提法下，创造诗情画意的造园事业中，我当年的"谬举"又为人所颂了。"含泪中的微笑",在我第一本书中,有着这样不平凡的经历啊!

我从这第一本书后,虽然留下过一点"疮痕",但我并没有气馁,我仍坚持着我的写作,到如今更有了新的发展。在这里我体会到,对一个为学的人来说,毅力是最大的动力;世界上没有平坦的道路,方向正确后,在于你有没有勇气走。如今我每见到这本《苏州园林》,我总是别有一番滋味,"我有柔情忘未了,卅年恩怨尽苏州"。我想这样来讲,我的感情还是真实的。

近三十年的年华过去了,我也垂垂老矣,然而"天意怜幽草,人间爱晚晴"。我还应该继续发挥余热,能为社会主义祖国文化事业,再写几本书,我这样期望着。

原载于一九八五年一月五日《书讯报》

In Lieu of Preface, My First Book
Suzhou Gardens
Chen Congzhou

To be honest, my first book should have been my earliest book. However I am a person with a wide range of interest. My earliest books are not on the line of my profession, for example, the *Biography of Poet Xu Zhimo*（徐志摩年谱）. This biographical writing is the brainchild of an emotional impulse. Besides, I wrote miscellaneous monographs on architecture. They are based on some materials I collected occasionally. The first book I wrote in real earnest should be the *Suzhou Gardens*. It was completed in 1956. This is the first book on a systematic study and discussion of Suzhou gardens after 1949 when the People's Republic of China was established.

In early nineteen fifties, I was a teacher in the Architecture Department of Tong Ji University in Shanghai（上海同济大学）. At the same time I did some spare-time teaching at Su Nan Industrial College in Suzhou（苏州苏南工专）. These classes were scheduled on Saturday morning. I took the night train on Fridays, stayed in a hotel near Guan Qian Jie（观前街）. Earlier next morning I went to the college located at Cang Lang Pavilion（沧浪亭）to teach my class. After taking a siesta, I rambled in the gardens, with notebook, camera, tape measure and scratch paper at hand. "I watch leisurely mosses and berries, as my migrant interest dictates." I enjoyed the sceneries and recorded their beauty at will. On early Sunday morning, I entered the gardens again. At noon, with a pot of fragrant tea and a dish of pastries, I sat by the rockeries to savor my snack. At that time, there were few viewers in these gardens. I could linger around as long as I wish. Excited by the scenes, I was not aware of the drowsiness. It was not until sunset that I left gardens with full satisfaction, and took the train back to Shanghai. My kids were waiting for me at the door and were elated with the Suzhou delicacies I brought back. Now they are as old as I was at that time.

It is already the era of my next generation when the second edition of my book *Suzhou Gardens* came to light in Japan the year before last.

I did the same every week for the next few years. I derived great pleasure from the excursions. Then I started to write the book based on the material and the thoughts I accumulated during this period. My viewpoints presented in the book are abstracted from the factual materials I have collected, and so, I believe, are fully substantiated. Touring the gardens triggered noble sentiment. Chinese gardens in essence catered to artists and men of letters. Therefore, the design and construction are underlined by poetic grace and picturesque romance. So naturally when I sorted out and edited the photographs I had taken of the gardens, I recalled those sentences and phrases from our ancients' poems. I entitled each photo with words or phrases from Song ci. As a result, I unwittingly turned a professional book on garden architecture into a literary writing. This seemed an act of originality on my part. After it was published, the book was highly acclaimed by the readers. But later in 1958, to my surprise, I became a target of criticism for the so called feudalistic scholar's ideas in the book. I had no way out but to bow to these accusations against the alleged outworn concept of mine. Almost 30 years later, with the new-emerging advocacy of closer integration of science with humanities, garden construction began to be imbued with artistic imagination and poetic inspiration. My much depreciated "absurdities" of the past was again highly appreciated by the public. "Smiling with tears" is what my first book offers me.

My first book left me with a bit of "trauma", but I have not been disheartened. In these years I have never relaxed my efforts to write and up till now made much headway. To a learner, perseverance is the most important motive force in life. There is no easy path in one's career. Sink or swim, all depends on whether one has courage to pursue the objective one has correctly chosen. I have a very peculiar aftertaste each time I pick up my first book. "Three decades of joy and woe all arise from Suzhou; On recalling it I'm still tender-hearted and full of yearning." My passion for Suzhou and its gardens is still smouldering, to tell you the truth.

I am gaining on age now. However, "The Heaven pities the somber grass; The Earth cherishes the golden twilight." I expect to spend the remainder of my life writing more books in dedication to my beloved people and country.

The original from *News of Books* 《书讯报》 of January 5th, 1985

苏州园林初步分析

Analysis of Suzhou Gardens

(in Chinese)

苏州园林初步分析

一

我国园林，如从历史上溯源的话，当推古代的囿与园，以及《汉制考》上所称的苑。《周礼·天官·大宰》："九职二曰园圃，毓草木。"《地官·囿人》："掌囿游之兽禁，牧百兽。"《地官·载师》："以场圃任园地。"《说文》："囿，苑有垣也。一曰禽兽曰囿。圃，种菜曰圃。园，所以树果也。苑，所以养禽兽也。"据此则囿、园、苑的含义已明。我们知道豨韦的囿，黄帝的圃已开囿圃之端，到了三代苑囿专为狩猎的地方，例如周姬昌（文王）的囿，刍荛雉兔，与民同利。秦汉以后，园林渐渐变为统治者游乐的地方，兴建楼馆，藻饰华丽了。秦嬴政（始皇）筑秦宫，跨渭水南北，覆压三百余里。汉刘彻（武帝）营上林苑，甘泉苑，以及建章宫北的太液池，在历史的记载上都是范围很大的。其后刘武（梁孝王）的兔园，开了叠山的先河。魏曹丕（文帝）更有芳林园。隋杨广（炀帝）造西苑。唐李漼（懿宗）于苑中造山植木，建为园林。北宋赵佶（徽宗）之营艮岳，为中国园林之最著于史籍者。宋室南渡，于临安（杭州）建造玉津、聚景、集芳等园。元忽必烈（世祖）因辽金琼华岛为万岁山太液池。明清以降除踵前遗规外，并营建西苑、南苑，以及西郊畅春、清漪、圆明等诸园，其数目视前代更多了。

私家园林的发展，汉代袁广汉于洛阳北邙山下筑园，东西四里，南北五里，构石为山，复畜禽兽其间，可见其规模之大了。梁冀多规苑囿，西至弘农，东至荥阳，南入鲁阳，北到河淇，周回千里。又司农张伦造景阳山，其园林布置有若自然。可见当时园林在建筑艺术上已有很高的造诣了。尚有茹皓，吴人，采北邙及南山佳石，复筑楼馆列于上下，并引泉莳花，这些都是以人工来代天巧。魏晋六朝这个时期，是中国思想上起一个大转变的时代，而亦中国历史上战争最频繁的时代，士大夫习于服食，崇尚清谈，再兼以佛学昌盛，于是礼佛养性，遂萌出世之念，虽居城市，辄作山林之想。在文学方面有咏大自然的诗文，绘画方面有山水画的出现，在建筑方面就在第宅之旁筑园了。石崇在洛阳建金谷园，从其《思归引序》来看其设计主导思想，是"避嚣烦"，"寄情赏"。再从《梁书·萧统传》，徐勉《诫子嵩书》，庾信《小园赋》等来看，他们的言论亦不外此意。唐代如宋之问的蓝田别墅，李德裕的平泉别墅，王维的辋川别业，皆有竹洲花坞之胜，清流翠篆之趣，人工景物，仿佛天成。而白居易的草堂，尤能利用自然，参合借景的方法。宋代李格非《洛阳名园记》，周密《吴兴园林记》，前者记北宋时所存隋唐以来洛阳名园如富郑公园等，后者记南宋吴兴园林如沈尚书园等，记中所述，几与今日所见园林无甚二致。明清以后，园林数目远迈前代，如北京勺园、漫园，扬州影园、九峰园、马氏玲珑馆，海宁安澜园，杭州小有天园等，以及明王世贞《游金陵诸园记》所记东园等诸园，其数已不胜枚举。今日存者如杭州皋园，南浔适园、宜园、小莲庄，上海豫园，常熟燕园，南翔古猗园，无锡寄畅园等，为数尚多，而苏州一隅又为各地之冠，如今我们来看看苏州园林在历史上的发展。

二

苏州在政治经济文化上，远在春秋时的吴，已经有了基础，其后两汉两晋逐渐发展。春秋时吴之梧桐园，以及晋之顾辟疆园，已开苏州园林的先声。六朝时江南已为全国富庶之区，扬州、南京、苏州等处的经济基础，到后来形成有以商业为主，有以丝织品及手工业为主，有为官僚地主的消费城市。苏州就是手工业重要产地兼官僚地主的消费城市。

我们知道六朝以还，继以隋唐，杨广（炀帝）开运河，促使南北物资交流，唐以来因海外贸易，江南富庶视前更形繁荣。唐末中原诸省战争频繁，受到很大的破坏，可是南唐吴越所在范围，在政治上经济上尚是小康局面，因此有余力兴建园林，宋时苏州朱长文因吴越钱氏旧园而筑乐圃，即是一例。北宋江南上承南唐吴越之旧，地方未受干戈，经济上没有受重大影响，园林兴建未辍。及赵构（高宗）南渡，苏州又为平江府治所在，赵构曾一度"驻跸"于此，王唤营平江府治，其北部凿池构亭，即官衙亦附以园林。其时土地兼并已甚，豪门巨富之宅，园林建筑不言可知了。故两宋之时，苏州园林著名者，如苏舜钦就吴越钱氏故园为沧浪亭，梅宣义构五亩园，朱长文筑乐圃，而朱勔为赵佶营艮岳外，复自营同乐园，皆较为著名。元时江浙仍为财富集中之地，故园林亦有所兴建，如狮子林，即其一例。迨入明清，土地兼并之风更甚，而苏州自唐宋以来是丝织品与各种美术工业品的产地，又为地主官僚的集中地，并且由科举登第者最多，以清一代而论，状元数字之多为全国冠，这些人年老归家，购田宅，设巨肆，除直接从土地上剥削外，再从商业上经营盘剥，以其所得大建园林以娱晚境。而手工业所生产，亦供若辈使用。其经济情况大略如此。它与隋唐洛阳，南宋吴兴，明代南京，是同样情况的。

除了上述情况之外，在自然环境上，苏州水道纵横，湖泊罗布，随处可得泉引水。兼以土地肥沃，花卉树木易于繁滋。当地产石，除尧峰山外，洞庭东西二山所产湖石，取材便利。距苏州稍远的如江阴黄山，宜兴张公洞，镇江圌山、大岘山，句容龙潭，南京青龙山，昆山马鞍山等所产，虽不及苏州为佳，然运材亦便。而苏州诸园之选峰择石，首推湖石，因其姿态入画，具备造园条件。《宋书·戴颙传》："山居吴下，士人共为筑室，聚石引水植林开涧，少时繁密，有若自然"，即其一例。其次苏州为人文荟萃之所，诗文书画人才辈出，士大夫除自出新意外，复利用了很多门客，如《吴风录》载："朱勔子孙居虎丘之麓，以种艺选石为业，游于王侯之门，俗称花园子。"又周密《癸辛杂识》云："工人特出吴兴,谓之山匠,或亦朱勔之遗风。"既有人为之策划，又兼有巧匠，故自宋以来造园家如俞澂、陆叠山、计成、文震亨、张涟、张然、叶洮、李渔、仇好石、戈裕良等，皆江浙人。今日叠石匠师出南京、苏州、金华三地，而以苏州匠师为首，是有历史根源的。但士大夫固然有财力兴建园林，然《吴风录》所载："虽闾阎下户亦饰小山盆岛为玩"，这可说明当地人民对自然的爱好了。

苏州园林在今日保存者为数最多，且亦最完整，如能全部加以整理，不啻是个花园城市。故言中国园林，当推苏州了，何怪大家都说"江南园林甲天下，苏州园林甲江南"的光荣称号呢。这些园林我经过五年的调查踏勘，复曾参与修复工作，前夏与今夏又率领同济大学建筑系同学作教学实习，主要对象是古建筑与园林，逗留时间较久，遂以测绘与摄影所得，利用拙政园、留园两个最大的园作例，略略加以说明一些苏州园林在历史上的发展，与设计方面的手法，供大家研究，至于其他的一些小园林有必要述及的，亦一并包括在内。图录内所附加词句，系集宋词，对景物仅略作点缀而已。

三

拙政园：拙政园在娄、齐二门间的东北街。明嘉靖中（公元1522年—1566年）王献臣因大宏寺废地营别墅，是此园的开始。"拙政"二字的由来，是用潘岳"拙者之为政"的意思。后其子以赌博负失，归里中徐氏。清初属海宁陈之遴，陈因罪充军塞外，此园一度为驻防将军府，其后又为兵备道馆。吴三桂婿王永宁亦曾就居于此园。后没入公家，康熙初改苏松常道新署，其时玄烨（康熙）南巡，也来游到此。苏松常道缺裁，散为民居。乾隆初归蒋棨，易名复园。嘉庆中再归海宁查世倓，复归平湖吴菘圃。迨太平天国克复苏州又为忠王府的一部分，太平天国失败，为反动统治所据，同治十年（公元1871年）改为八旗奉直会馆，仍名拙政园，西部归张履谦所有，易名补园，解放后已合而为一。

拙政园的布局主题是以水为中心，池水面积约占总面积五分之三，主要建筑物十之八九皆临水而筑（图275页，拙政园平面图），文徵明《拙政园记》："郡城东北界娄、齐门之间，居多隙地，有积水亘其中，稍加浚治，环以林木。……"据此可以知道是利用原来地形而设计的。与明末计成《园冶》中《相地》一节所说"高方欲就亭台，低凹可开池沼……"的因地制宜方法相符合的。故该园以水为主，实有其道理在。在苏州不但此园如此，阔阶头巷的网师园，水占全园面积达五分之四。平门的五亩园亦池沼逶迤，望之弥然，莫不利用原来的地形而加以浚治的。景德路环秀山庄乾隆间蒋楫凿池得泉，名"飞雪"，亦是解决水源的好办法。

园可分中、西、东三部，中部系该园主要部分，旧时规模所存尚多，西部即张氏补园，已大加改建，然布置尚是平妥。东部为明王心一归田园居，久废，正在重建中。

中部远香堂为该园的主要建筑物（图86页），单檐歇山面阔三间的四面厅，从厅内通过窗棂四望，南为小池假山，广玉兰数竿，扶疏接叶，云墙下古榆依石，幽竹傍岩，山旁修廊曲折，导游者自园外入内。似此的布置不但在进门处可以如入山林（图84页上），而坐厅南望亦有山如屏，不觉有显明的入口，它与住宅入口处置内照壁，或置屏风等来作间隔的方法，采用同一的手法。东望绣绮亭（图90页），西接倚玉轩（图 86 页），北临荷池，而隔岸雪香云蔚亭（图 105 页）与待霜亭（图 104 页）突出水面小山之上。游者坐此厅中，则一园之景可先窥其轮廓了。以此厅为中心的南北轴线上，高低起伏，主题突出。而尤以池中岛屿，环以流水，掩以丛竹，临水湖石参差，使人望去殊多不尽之意，仿佛置身于天然池沼中。从远香堂缘水东行跨倚虹桥（图 96 页），桥与阑皆甚低，系明代旧构。越桥达倚虹亭，亭倚墙而作，仅三面临空，故又名东半亭。向北达梧竹幽居（图 102 页），亭四角攒尖，每面辟一圆拱门。此处系中部东尽头，通过二道圆拱门望池中景物，如入环中（图 99 页）。而隔岸极远处的西半亭隐然在望，是亭内又为一圆拱门，倒映水中，所谓别有洞天以通西部的，亭背则北寺塔耸立云霄中，为极妙的借景。左顾远香堂、倚玉轩及香洲等，右盼两岛，前者为华丽的建筑群，后者为天然图画（图 97 页）。刘师敦桢云："此为园林设计上运用最好的对比方法。"实际而言，东西两岸水面距离并不太大，然看去反觉深远特甚，因设计时在水面隔以梁式石桥，逶迤曲折，人们视线从水面上通过石桥才达彼岸，加以两旁一面是人工华丽的建筑，一面为天然苍翠的小山，二者之间水是修长的，自然给人们的感觉是更加深远与扩大了。而对岸老榆傍岸，垂杨临水，其间一洞窈然，楼台画出，又别有天地了。从梧竹幽居经三曲桥，小径分歧，屈曲循道登山，达巅为待霜亭（图 104 页），亭六角，翼然出丛竹间，向东襟带绿漪亭（图 100 页），西则复与长方形的雪香云蔚亭相呼应（图 105 页），此岛平面为三角形，与雪香云蔚亭一岛椭圆形者有别，二者之间一溪相隔，溪上覆以小桥（图 108 页），其旁幽篁丛出，老树斜倚，而清流涓涓宛若与树上流莺相酬答。至此顿忘尘嚣了。自雪香云蔚亭下，便到荷风四面亭（图 125 页），亭亦六角，居三路之交点，前后皆以曲桥相贯，前通倚玉轩而后达见山楼及别有洞天。经曲廊名柳阴路曲者（图 116 页）达见山楼（图 109 页），楼为重檐歇山顶，以假山构成云梯可导至楼层，是楼地位居中部西北之角，因此登楼远望，其至四周距离较大，所见景物亦远，如转眼北眺，则城隈景物，又瞬入眼帘了。此种手法，在中国园林中最为常用，如中由吉巷半园用五边形亭，狮子林用扇面亭（图 206 页），皆置于角间略高的山巅。至于此园面积较大，而积水弥漫，建一重楼，但望去不觉高耸凌云，而水间倒影清澈，尤增园林景色。然在设计时应注意其立面线脚，宜多用横线，盖与水面取得平行，以求统一。香洲俗呼旱船，形似船而不能行水者（图117、118页），入舱置一大镜，故从倚玉轩西望，镜中景物，真幻莫辨。楼上名澂观楼，亦宜眺远。向南为得真亭（图123页），内置一镜，命意与前同。是区水面狭长，上跨石桥名小飞虹（图120页），将水面划分为二，其南水榭三间，名小沧浪，亦跨水上，又将水面再度划分，二者之下皆空，不但不觉其局促，反觉面积扩大，空灵异常，层次渐多了。人们视线从小沧浪穿小飞虹及一庭秋月啸松风亭（图120、122页），水面极为辽阔，而荷风四面亭倒影、香洲侧影、见山楼角皆先后入眼中（图121页），真有从小窥大，顿觉开朗的样子。枇杷园在远香堂东南，以云墙相隔，通月门（图87页），则嘉实亭与玲珑馆分列于前，复自月门回望雪香云蔚亭，如在环中，此为最好的对景（图88页）。我们坐园中，垣外高槐亭台，移置身前，为极好的借景。园内用鹅子石铺地，雅洁异常，惜沿墙假山修时已变更原形，而云墙上部无收头，转折又略嫌生硬（图87页）。从玲珑馆旁曲廊至海棠春坞（图92页），屋仅面阔二间（图93页），阶前古树一木，海棠一树，佳石一二，近屋以短廊，漏窗外亭阁水石，隐约在望（图94、95页），其环境表面上看来是封闭的，而实际是处处通畅，面面玲珑，置身其间，便感到密处有疏，小处现大，可见设计手法运用的巧妙了。

西部与中部原来是不分开的，后来一园划分为二，始用墙间隔，如今又合而为一，因此墙上开了漏窗。当其划分时，西部欲求有整体性，于是不得不在小范围内加工，沿水的墙边就构了水廊，廊复有曲折高低的变化，人行其上，宛若凌波（图132页），是苏州诸园中之游廊极则。三十六鸳鸯馆与十八曼陀罗花馆系鸳鸯厅（图141页），为西部主要建筑物，外观为歇山顶，面阔三间，内用卷棚四卷（图317页，三十六鸳鸯馆及十八曼陀罗花馆横断面图），四隅各加暖阁，其形制为国内唯一孤例。此厅体积似乎较大，其因实由于西部划分后，欲成为独立的单位，此厅遂为主要建筑部分，在需要上不能不建造，但碍于地形，于是将前部空间缩小，后部挑出水中，虽然解决了地位安顿问题，但卒使水面变狭与对岸之山距离太近，陆地缩小，而本身又觉与全园不称，当然是美中不足处。此厅为主人宴会与顾曲之处，因此在房屋结构上除运用卷棚顶以增加演奏效果外，其四隅之暖阁，既解决进出时风击问题，复可利用为宴会时仆从听候之处，演奏时暂作后台之用，设想上是相当周到，内部的装修精致与留听阁同为苏州少见的。至于初春十八曼陀罗花馆看宝朱山茶花，夏日三十六鸳鸯馆看鸳鸯于荷蕖间，宜乎南北各置一厅了。对岸为浮翠阁，八角二层，登阁可鸟瞰全园，惜太高峻，与环境不称。其下隔溪小山上置二亭，即笠亭（图144页）与扇面亭（图145页），亭皆不大，盖山较低小，不得不使然。扇亭位于临流转角，因地而设，宜于闲眺，故颜其额为"与谁同坐轩"。亭下为修长流水，水廊缘边以达倒影楼（图135页），楼为歇山顶，高二层，与六角攒尖的宜两亭遥遥相对（图137页），皆倒影水中，互为对景。鸳鸯厅西部之溪流中置塔影亭（图149页），它与其北的留听阁（图148页），同样在狭长的水面二尽头，而外观形式亦相仿佛，不过地位视前二者为低，布局与命意还是相同的。塔影亭南原为补园入口以通张宅的，今已封闭。

东部久废，刻在重建中，从略。

留园：在阊门外留园路，明中叶为徐泰时东园，清嘉庆间（约公元1800年左右）刘恕重建，称寒碧山庄，又名刘园。园中旧有十二峰，为太湖石之上选。光绪二年（公元1876年）间归盛康，易名留园。园占地五十市亩，面积为苏州诸园之冠（图325页）。

是园可划分为东西中北四部，中部以水为主，环绕山石楼阁，贯以长廊小桥。东部以建筑为主，列大型厅堂，参置轩斋，间列立峰斧劈，在平面上曲折多变。西部以大假山为主，漫山枫林，亭榭一二，南面环以曲水，仿晋人武陵桃源，是区与中部以云墙相隔，红叶出粉墙之上，望之若云霞，为中部最好的借景。北部旧构已毁，今又重辟，平淡无足观，从略。

中部：入园门经二小院至绿荫（图155页），自漏窗北望，隐约见山池楼阁片断，向西达涵碧山房三间（图157页），硬山造，为中部的主要建筑，前为小院，中置牡丹台，后临荷池。其左明瑟楼倚涵碧山房而筑（图159页），高二层屋顶用单面歇山（图333页上，明瑟楼正立面图），外观玲珑，由云梯可导至二层。复从涵碧山房西折上爬山游廊，登闻木樨香轩（图163页），坐此可周视中部，尤其东部曲谿楼，西楼，清风池馆，汲古得绠处及远翠阁等，参差前后，高下相呼的诸楼阁掩映于古木奇石之间。南面则廊屋花墙，水阁联续，而明瑟楼微突水面，涵碧山房之凉台再突水面，层层布局，略作环抱之势，楼前清水一池，倒影历历在目。自闻木樨香轩向北东折经游廊达远翠阁，是阁位置于中部东北角（图167页），其用意与拙政园见山楼相同，不过一在水一在陆，又紧依东部，隔花墙为东部最好的借景（图168页）。小蓬莱宛在水中央，濠濮亭列其旁（图169页），皆几与水平，如此对比，容易显山之峻与楼之高了。曲豀楼底层西墙皆列砖框（图170页），漏窗，游者至此感觉处处邻虚，移步换影，眼底如画了。而尤其举首西望，秋时枫林如醉，衬托于云墙之后，其下高低起伏若波然，最令人依恋不已。北面为假山，可亭六角出假山之上（图162页），其后则为长廊了（图166页）。

东部主要建筑物有二：其一五峰仙馆（楠木厅），面阔五间，系硬山造（图173页），内部装修陈设，精致雅洁，为江南旧式厅堂布置之上选（图174页）。其前后左右皆有大小不等的院子，前后二院皆列假山，人坐厅中仿佛面对岩壑，然此法为明计成所不取，《园冶》云："人皆厅前掇山，环堵中耸起高高三峰，排列于前，殊为可笑。"此厅列五峰于前，似觉太挤，了无生趣。而计成认为在这种情况下应该是"以予见或有嘉树稍点玲珑石块，不然墙中嵌埋壁岩，或顶植卉木垂萝，似有深境也。"我觉得这办法是比较妥善多了。后部小山前有清泉一泓，境界至静，惜源头久没，泉呈时涸时有之态。山后沿墙绕以回廊，可通左右前后，游者至此，偶一不慎，方向莫辨。在此小院中左眺远翠阁，则隔院楼台又炯然在目（图168页）。使人益觉该园之宽大了。其旁汲古得绠处，小屋一间，紧依五峰仙馆，入内则四壁皆虚，中部景物又复现眼前。其与五峰仙馆相联接处的小院，中植梧桐一树，望之亭亭如盖，此小空间的处理是极好的手法。还我读书处与揖峰轩都是两个小院，在五峰仙馆的左邻，是介于与林泉耆硕之馆中间，为两大建筑物中之过渡。小院绕以回廊，间以砖框，院中安排佳木修竹，萱草片石，都是方寸得宜，楚楚有致，使人有静中生趣之感，充分发挥了小院落的设计手法（图178-183页）。而游者至此往往相失。由揖峰轩向东为林泉耆硕之馆，俗呼鸳鸯厅，装修陈设极尽富丽（图184页），屋面阔五间，单檐歇山造，前后二厅，内部各施卷棚（图349、350页，林泉耆硕之馆正立面图，纵断面图，侧立面图，和横断面图），主要一面向北，大木梁架用"扁作"，有雕刻，南面用"圆作"，无雕刻。厅北对冠云沼，冠云（图185页），岫云（图190页下），瑞云（图191页）三峰以及冠云亭、冠云楼。三峰为明代旧物，苏州最大的湖石。冠云峰后侧为冠云亭（图185、187页），亭六角，倚玉兰花下。向北登云梯上冠云楼，虎丘塔影，阡陌平畴，移置窗前了。伫云庵与冠云台位于沼之东西。从冠云台入月门乃佳晴喜雨快雪之亭，亭内楠木榻扇六扇，雕刻甚精，为吴中装修之极品（图192页），惜是亭面西，难免受阳光风露之损伤。东园一角系新辟，山石平淡无奇，不足与旧构相颉颃了。

西部园林以时代而论，似为明东园旧规，山用积土，间列黄石，犹是李渔所云"小山用石，大山用土"的老办法，因此漫山枫树，得以滋根，林中配二亭；一为舒啸亭（图193页）系圆攒尖，一为至乐亭（图194页）六边形，系仿天平山范祠御碑亭而略变形的，在苏南还是创见。前者隐于枫林间，后者据西北山腰，可以上下眺望。南环清溪，植桃柳成荫，原期使人至此有世外之感，但有意为之，顿成做作，以人工胜天然在园林中实是不易的事。溪流终点，则为活泼泼地（图197页），一阁临水，水自阁下流入，人在阁中，仿佛跨溪之上，不觉有尽头了。唯该区假山，经数度增修，殊失原态。

北部旧构已毁，今新建，无亭台花木之胜。

四

江南园林占地不广，然千岩万壑，清流碧潭，皆宛然如画，正钱泳所说："造园如作诗文，必使曲折有法。"因此对于山水、亭台、厅堂、楼阁、曲池、方沼、花墙、游廊等之安排划分，必使风花雪月，光景常新，不落窠臼，始为上品。因此对于总体布局及空间处理，务使有扩大之感，观之不尽，而风景多变，极尽规划的能事。总体布局可分以下几种。

中部以水为主题，贯以小桥，绕以游廊，间列亭台楼阁，大者中列岛屿，此类如网师园，以及怡园等之中部。庙堂巷畅园，地颇狭小，一水居中，绕以廊屋，宛如盆景。

留园虽以水为主，然刘师敦桢认为该园以整体而论，当以东部建筑群为主，这话亦有其理存在着。

以山石为全园之主题，因是区无水源可得，且无洼地可利用，故不能不以山石为主题使其突出，固设计中一法，西百花巷程氏园无水可托，不得不如此了。环秀山庄范围小，不能凿大池，亦以山石为主，略引水泉，俾山有生机，岩现活态，苔痕鲜润，草木华滋，宛然若真山水了。

基地积水弥漫，而占地尤广，布置遂较自由，不能为定法所囿，如拙政园、五亩园等较大的，更能发挥开朗变化的能事，尤其拙政园中部的一些小山大有张涟所云："平冈小坡，曲岸回沙。"都是运用人工方法来符合自然的境界。计成《园冶》云："虽由人作，宛自天开。"刘师敦桢主张："池水以聚为主，以分为辅，小园聚胜于分，大园虽可分，但须宾主分明。"网师园与拙政园是两个佳例，皆苏州园林上品。

前水后山，复构堂于水前，坐堂中穿水遥对山石，而堂则若水榭，横卧波面，文衙弄艺圃布局即如是。北寺塔东芳草园亦仿佛似之。

中列山水，四周环以楼及廊屋，高低错落，迤逦相续，与中部山石相呼应，如小新桥巷耦园东部，在苏州尚不多见。东北街韩氏小园，亦略取是法，不过楼屋仅有两面。中由吉巷半园，修仙巷宋氏园皆有一面用楼。

明代园林，接近自然，犹是计成、张涟辈后来所总结的方法，利用原有地形，略加整理，其所用石，在苏州大体以黄石为主，如拙政园中部二小山及绣绮亭亭下的，黄石虽无湖石玲珑剔透，然掇石有法，反觉浑成，既无矫揉做作之态，且无累石不固的危险。我们能从这种方法中细细探讨，在今日造园中还有不少优良传统可以吸收学习的。到清代造园率皆以湖石叠砌，贪多好奇，每以湖石之多少与一峰之优劣，与他园计较短长，试以怡园而论，购洞庭山三处废园之石累积而成，一峰一石，自有上选，但其西北部之一区，石骨外露，卒无活态，即其一例。可见"小山用石"，非全无寸土，不然树木将无所依托了。环秀山庄虽改建于乾隆间，数弓之地，深溪幽壑，势若天成，运用宋人山水的所谓"斧劈法"，再以镶嵌出之，简洁遒劲（图217页），其水则迂回曲折，山石处处滋润，苍岩欣欣欲活了，诚为江南园林的杰构。于此方知设计者若非胸有丘壑，挥洒自如者焉能至斯？学养之功可见重要了。

掇山既须以原有地形为据，但自然之态，变化多端，无一定成法，可是自然的形成与发展，亦有一定规律可循，"师古人不如师造化"，实有其理在。我们今日能通乎此理。从自然景物加以分析，证以古人作品，评其妍媸，撷其菁华，构成最美丽的典型。奈何苏州所见晚期园林，什九已成"程式化"，从不在整体考虑，每以亭台池馆，妄加拼凑，尤以掇山选石，皆举一峰片石，视之为古董，对于花树的衬托，建筑物的调和等，则有所忽略，这是今日园林设计者要引以为鉴的。如怡园欲集诸园之长，但全局涣散，似未见成功。

园林之水，首在寻源，因无源之水必死水。如拙政园利用原来池沼。环秀山庄掘地得泉，水虽涓涓，亦必清洌可爱。但园林面积既小，欲使有汪洋之概，则在设计时的运用，其法有二：一、池面利用不规则的平面，间列岛屿，上贯以小桥，在空间上使人望去，不觉一览无余。二、留心曲岸水口的设计，故意做成许多湾头，望之仿佛有许多源流，如是则水来去无尽头，有深壑藏幽之感。至于曲岸水口之利用芦苇，杂以菰蒲，则更显得隐约迷离，这是以较大的园林应用才妙。留园活泼泼地，水榭临流，溪至榭下已尽，但必流入一部分，则俯视之下，榭若跨溪上，水不觉终止。南显子巷惠荫园水假山，系层叠巧石如洞曲，引水灌之，点以步石，人行其间，如入涧壑，洞上则构屋。此种形式为吴中别具一格者，殆系南宋杭州赵翼王园中之遗制。沧浪亭以山为主，但东部的步廊突然逐渐加高，高瞰水潭，自然临渊莫测。惜是潭平时无涓涓活流，致设计者事与愿违，斯亦不能不注意的。艺圃的桥与水几平（图215页），反之两岸山石愈显高峻了。怡园之桥虽低于山，但与水尚有一些距离，还不失为有所依据的。至于小溪作桥，在对比之下其情况何如？不难想像。古人遂改用"点其步石"的方法，则更为自然有致了。瀑布除环秀山庄檐瀑外，他则罕有。

中国园林除水石池沼外，建筑物如厅、堂、斋、台、亭、榭、轩、卷、廊等，都是构成园林的主要部分，然江南园林以幽静雅淡为主，故建筑物务求轻巧，方始相称，所在建筑物的地点，平面，以及外观不能不注意。《园冶》云："凡园圃立基，定厅堂为主，先取乎景，妙在朝南，倘有乔木数株，仅就中庭一二。"苏南园林尚守是法，如拙政园远香堂，留园涵碧山房等皆是。至于楼台亭阁的地位，虽无成法，但"按基形式"，"格式随宜"，"随方制象，各有所宜"，"一櫺一角，必令出自己裁"，"花间隐榭，水际安亭"，还是在设计人从整体出发而加以灵活应用。古代如《园冶》，《长物志》，《工段营造录》等，虽有述及，最后亦指出我们不能守为成法的。试以拙政园而论，我们自高处俯视，建筑物虽然是随宜安排的，其方向差不多都依地形而改变。其外观给人的感觉是轻快为主，平面正方形，长方形，多边形，圆形等皆有，屋顶形式则有歇山、单面歇山、硬山、悬山、攒尖等，而无庑殿式，即歇山、硬山、悬山，亦多数采用卷棚式。其翼角起翘类多用"水戗发戗"的办法，因此翼角起翘低而外观轻快。下玲珑的挂落，柱间微弯的吴王靠，得能取得一致。建筑物在立面的处理，以留园中部而论，我们自闻木樨香轩东望，对景主要建筑物是曲谿楼，用歇山顶，其外观在第一层做成仿佛台基的形状，与水相平行的线脚与上层分界，虽系二层看去不觉其高耸了。而尤其曲谿楼、西楼、清风池馆，三者的位置各有前后，屋顶立面皆同中寓不同。与下部的立峰水石都很相称，古木一树斜横波上，益增苍古，而墙上的砖框漏窗，上层的窗棂与墙面虚实的对比，疏淡的花影，都是苏州园林特有的手法，尤其倒影水中，其景更美。明瑟楼与涵碧山房相邻，前者为卷棚歇山，后者为卷棚硬山，然两者相联，不能不用变通的办法，明瑟楼歇山山面仅作一面，另一面用垂脊，不但不觉得其难看，反觉生动有变化（图159页）。他如畅园因基地较狭长，中又系水池，水榭无法安排，卒用单面歇山，实系同出一法。反之东园一角亭，为求轻巧起见，六角攒尖顶翼角用"水戗发戗"，其上部又太重，柱身瘦而高，在整个比例上顿觉不稳。东部舒啸亭、至乐亭，前者小而不见玲珑，后者屋顶虽多变化亦觉过重，都是比例上的缺陷。苏南筑亭，晚近香山匠师每将屋顶提得过高，但柱身又细，整个外观未必真美。反视艺圃明代遗构，屋顶略低，较平稳得多。终之单体建筑，必然要考虑到与全园的整个关系才是。至于平面变化，虽洞房曲户，亦必做到曲处有通，实处有疏。小型轩馆，一间二间，或二间半均可，皆视基地，随宜安排，如拙政园海棠春坞面阔二间，一大一小，宾主分明。留园揖峰轩面阔二间半，而尤妙于半间，方信《园冶》所云有所独见之处。建筑物的高下得势，左右呼应，虚实对比，在在都须留意。王洗马巷万氏园（原为任氏），园虽小，书房部分自成一区，极为幽静。其装修与铁瓶巷任宅东西花厅、顾宅花厅、网师园、西百花巷程氏园、大石头巷吴宅花厅等（详见拙著《装修集录》）都是苏州园林中之上选。至于他园尚多商量处，如留园太繁琐伧俗，佳者甚少；拙政园精者固有，但多数又觉简单无变化，力求一律，皆修理中东拼西凑或因陋就简所造成。怡园旧装修几不存，而旱船为吴中之尤者（图229页），所遗装修极精。

园林游廊为园林的脉络，在园林建筑中处极重的地位，故特地说明一下。今日苏州园林廊之常见者为复廊，廊系两面游廊中隔以粉墙，间以漏窗（详见拙编《漏窗》），使墙内外皆可行走，此种廊大都用于不封闭性的园林，如沧浪亭的沿河（图261页）。或一园中须加以间隔，欲使空间扩大，并使入门有所过渡，如怡园的复廊，便是一例，此廊显然是仿前者。它除此作用外，因岁寒草堂与拜石轩之间不为西向的阳光与朔风所直射，用以阻之，而阳光通过漏窗，其图案更觉玲珑剔透（图224页）。游廊有陆上、水上之分，又有曲廊、直廊之别。但忌平直生硬，今日苏州诸园所见，过分求曲，则反觉生硬勉强，如留园中部北墙下的。至其下施以砖砌阑干，一无空虚之感，与上部挂落不称，柱夹砖中，僵直滞重，铁瓶巷任宅及拙政园西部水廊小榭，易以镂空之砖，似此较胜，拙政园旧时柳阴路曲，临水一面阑干用木制，另一面上安吴王靠，是有道理的（图128页）。水廊佳者，如拙政园西部的，不但有极佳的曲折，并有适当的坡度，诚如《园冶》所云的"浮廊可渡"，允称佳构（图132页）。尤其可取的就是曲处置湖石芭蕉（图133、134页），配以小榭，更觉有变化。爬山游廊在苏州园林中狮子林、留园、拙政园略点缀一二，大都是用于园林边墙部分。设计此种廊时应注意到坡度与山的高度问题，否则运用不当，顿成头重脚轻，上下不协调。在地形狭宽不同的情况下，可运用一面坡的，或一面坡与二面坡并用，如留园西部的。曲廊的曲处是留虚的好办法，随便点缀一些竹石、芭蕉，都是极妙的小景。李斗云："板上甃砖谓之响廊，随势曲折谓之游廊，愈折愈曲谓之曲廊，不曲者修廊，相向者对廊，通往来者走廊，容徘徊者步廊，入竹为竹廊，近水为水廊。花间偶出数尖，池北时来一角，或依悬崖，故作危槛，或跨红板，下可通舟，递迢于楼台亭榭之间，而轻好过之。廊贵有栏。廊之有阑，如美人服半臂，腰为之细，其上置板为飞来椅，亦名美人靠，其中广者为轩。"言之尤详，可资参考。今日复有廊外植芭蕉呼为蕉廊，植柳呼为柳廊，夏日人行其间，更觉翠色侵衣，溽暑全消了。冬日则阳光射入，温和可喜，用意至善。而古时以廊悬画称画廊，今日壁间嵌诗条石，都是极好的应用。

园林中水面之有桥，正陆路之有廊，重要可知，苏州园林习见之桥，一种为梁式石桥，可分直桥、九曲桥、五曲桥、三曲桥、弧形桥等，其位置有高于水面与岸相平的，有低于两岸浮于水面的。以时代而论，后者似较旧，我们今日在艺圃及无锡寄畅园、常熟诸园（图270页）所见的都是如此，怡园及已毁木渎严家花园的亦仿佛似之，不过略高于水面一点。旧时为什么如此设计呢？它所表现的效果有二：第一，桥与水平，则游者凌波而过，水益显汪洋，桥更觉其危了。第二，桥低则山石建筑愈形高峻，与丘壑楼台自然成强烈对比。无锡寄畅园假山用平岗，其后以惠山为借景，岗下幽谷间施以是式桥，诚能发挥明代园林设计之高度技术。今日梁式桥往往不照顾地形，不考虑本身大小，随便安置，实属非当。而尤其阑干高度、形式，都从不与全桥及环境作一番研究，甚至于半封建半殖民地的铁阑干都加了上去，如拙政园西部的。而上选者，如艺圃小桥，拙政园倚虹桥（图96页）都是。拙政园中部的三曲五曲之桥，阑干比例都还好，可惜桥本身略高一些，待霜亭与雪香云蔚亭二小山之间石桥，仅搁一石板，不施阑干，极尽自然质朴之意，亦佳构（图108页）。另一种为小型环洞桥，狮子林、网师园都有，以此二桥而论，前者不及后者为佳，因环洞桥不适宜建于水中部，水面既小，用此中阻，遂显庞大质实，略无空灵之感。后者建于东部水尽头，桥本身又小（图219页），从西东望，辽阔的水面中倒影玲珑，反之自桥西望，月门倒映水中，用意相同。中由吉巷半园，因地位狭小，将环洞变形，亦系出权宜之计。至于小溪，《园冶》所云"点其步石"的办法，尤能与自然相契合，实远胜架桥其上，可是此法今日差不多已成绝响了。

园林的路，《清闲供》云："门内有径，径欲曲。""室旁有路，路欲分。"今日我们在苏州园林所见，还能如此。拙政园中部道路犹守明时旧规，从原来地形出发，加以变化，主次分明，曲折有度。环秀山庄面积小，不能不略作纡盘，但亦能恰到好处，行者有引人入胜之慨。然狮子林、怡园的，故作曲折，使人莫之所从，既背自然之理，又多不尽人情。因此矫揉做作，与自然相距太远的安排，实在是不艺术的事。

　　铺地，在园林亦是一件重要的工作，不论庭前、曲径、主路，皆须极慎重考虑，今日苏州园林所见：有仄砖铺地，大都铺于主路，施工简单，拼凑图案自由。碎石地，用碎石仄铺，可用于主路小径庭前。上面间有用缸片点缀一些图案。或缸片仄铺间以瓷片的。用法同前。鹅子地，或鹅子间加磁片拼凑成各种图案，视上述的要细致雅洁得多，但其缺点是石隙间的泥土，每为雨水及人力所冲扫而逐渐减少，又复较易长小草，保养费事，是需要改进的。冰裂地则用于庭前，苏南的结构有二：其一即冰纹石块平置地面，如拙政园远香堂前的（图152页），颇饶自然之趣，然亦有不平稳的流弊。其一则水纹石间交接处皆对卯拼成，施工难而坚固，如留园涵碧山房前（图198页）、铁瓶巷顾宅花厅的（图241页），都是极工整。至于庭前踏跺用天然石垒，如拙政园远香堂（图152页）及留园五峰仙馆前的（图173页），皆饶野趣。

　　园林的墙，有乱石墙、磨砖墙、漏砖墙、白粉墙等数种，苏州今日所见以白粉墙为最多，外墙有上开瓦花窗（漏窗开在墙顶部的），内墙间开漏窗及砖框的，所谓粉墙花影，为人乐道。磨砖墙园内仅建筑物上酌用之，园门十之八九贴以水磨砖砌成图案的，如拙政园大门（图83页）。乱石墙只见于裙肩处。在上海南市薛家浜路旧宅中我曾见到冰裂纹上缀以梅花的，极精，似系明代旧物。西园以水花墙区分水面，亦别具一格（图252页）。

　　联对匾额在中国园林中正是如人之有须眉，是不能少的一件重要点缀品，苏州又为人文荟萃之区，当时园林建造复有文人画家的参预，从人工上构成诗情画意，将平时所见真山水，古人名迹，诗文歌诗所表达的美妙意境，撷其精华而总合之，加以突出，因此山林岩壑，一亭一榭，莫不用文学上极典雅美丽而适当的词句来形容它，使游者入其地，览景而生情文，这些文字亦就是这个环境中最恰当的文字代表。例如拙政园的远香堂与留听阁，同样是一个赏荷花的地方，前者出"香远益清"句，后者出"留得残荷听雨声"句。留园的闻木樨香轩，拙政园的海棠春坞，又是都根据这区所种的树木来命名的。游者至此，不期而然的能够出现许多文学艺术的好作品，这不能不说是中国园林的一个特色了。我希望今后在许多旧园林中，如果无封建意识的文字，仅就描写风景的，应该好好保存下来。苏州诸园皆有好的题词，而怡园诸联集宋词，更能曲尽其意，可惜已经大部分不存了，至于用材料，因园林风大，故十之八九用银杏木阴刻，填以石绿。或用木阴刻后髹漆敷色者亦有，不过色彩都是冷色。亦有用砖刻的，雅洁可爱，字体以篆隶行书为多，罕用正楷，取其古朴与自然。中国书画同源，本身是个艺术品，当然是会增加美观的。

树木之在园林，其重要不待细述，已所洞悉。江南园林面积小，且都属封闭性，四周绕以高垣，故对于培花植木，必须深究地位之阴阳，土地之高卑，树木发育之迟速，耐寒抗旱之性能，姿态之古拙与华滋，更重要的为布置的地位与树石的安排了。园林之假山与池沼皆真山水的缩影，因此树木的配置，不能任其自由发展。所栽植者必须体积不能过大，而姿态务求毕备，虬枝旁水，盘根依阿，景物遂形苍老，因此在选树之时尤须留意此端。宜乎李格非云："人力胜者少苍古。"今日苏州树木常见的，如拙政园大树用榆，留园中部多银杏，西部则漫山枫树。怡园面积小，故易以桂，松，及白皮松，尤以白皮松树虽小而姿态古拙，在小园中最是珍贵。他则杂以松、梅、棕树、黄杨，在发育上均较迟缓。其次园小垣高，阴地多而阳地少，于是墙阴必植耐寒植物，如女贞、棕树、竹之类，岩壑必植高山植物，如松、杉之类。阶下石隙之中，植长绿阴性草类，全园中长绿者遂多于落叶，则四季咸青，不致秋冬秃无物了。至于乔木若榆、槐、枫等，除每年修枝使其姿态古拙入画外，此种树的根部甚美，尤以榆树等树龄大后，身空皮留，老干抽条，葱翠如画境。今日苏州园林中山巅栽树，大别有两种情况：第一类山巅山麓只植大树而虚其根部，俾可欣赏其根部与山石之美，如留园的与拙政园的一部分。第二类山巅山麓树木皆出丛竹或灌木之上，山石并攀以藤萝，使望去有深郁之感，如沧浪亭及拙政园的一部分。然二者设计者的依据有所不同，以我的分析，这些全在设计者所用粉本的各异，如前者师元代画家倪瓒（云林）的清逸作风，后者则效明代画家沈周（石田）的沉郁了。至于滨河低卑之地，种柳，栽竹，植芦，而墙阴栽爬山虎、修竹、天竹、秋海棠等，叶翠、花冷、实鲜，高洁耐赏。但此等亦必须每年修剪，不能任其发育。

园林栽花与树木同一原则，背阴且能略受阳光之地，栽植桂花，山花之类，此二者除终年常青外，且花一在秋，一在春初，都是群花未放之时，而姿态亦佳，掩映于奇石之间，冷隽异常。紫藤则入春后一架绿阴，满树繁花，望之若珠光宝露。牡丹作台，衬以文石阑干，实牡丹宜高地向阳，兼以其花华丽，故不得不使然。他若玉兰、海棠、牡丹、桂花等同栽庭前，谐音为玉堂富贵，当然命意已不适于今日，但对于开花的季节与色彩的安排，前人未始无理由的。桃李则宜植林，适于远眺，此在苏州，仅范围大的如留园、拙政园可以酌用之。

　　树木的布置，在苏州园林有两个原则：第一，用同一种树植之成林，如怡园听涛处植松，留园西部植枫，闻木樨香轩前植桂。但又必须考虑到高低疏密间及与环境的关系。第二，用多种树同植，其配置如作画构图一样，更要注意树的方向及地的高卑是否适宜于多种树性，树叶色彩的调和对比，长绿树与落叶树的多少，开花季节的先后，树叶形态，树的姿势，树与石的关系，必须要做到片山多致，寸石生情，二者间是一个有机的联系才是。更须注意的它与建筑物的式样，颜色的衬托，是否已做到好花须映好楼台的效果。水中植荷，似不宜多，因荷多必减少了水的面积，楼台缺少倒影，故宜略点缀一二，亭亭玉立，摇曳生姿，隔秋水宛在水中央。据云昆山顾氏园藕植于池中石板底，石板仅凿数洞，俾不使其自由繁殖。刘师敦桢云："南京明徐氏东园池底置缸，植荷其内。"用意相同。

　　苏南园林以整体而论，其色彩以雅淡幽静为主，它与北方皇家园林的金碧辉煌，适成对比。其所以如此，以我个人见解：第一，苏南居住建筑，所施色彩在梁枋柱头皆用栗色，挂落用墨绿，有时柱头用黑色退光，都是一些冷色调，与白色墙面，起了强烈的对比，而花影扶疏，又适当地冲淡了墙面强白，形成良好的过渡，自多佳境了。且苏州园林皆与住宅相连，为养性读书之所，更应以清静为主，宜乎有此色调，它与北方皇家花园的那样宣扬自己威风与炫耀富贵的，在作风上有所不同。但苏州园林士大夫未始不欲炫耀富贵，然在装修、选石、陈列上用功夫，在色彩上仍然保持以雅淡为主的原则。再以南宗山水而论，水墨浅绛，略施淡彩，秀逸天成，早已印在士大夫及文人画家的脑海中，自然由这种思想影响下，设计出来的园林，当然不会用重彩贴金了。加以江南炎热，朱红等热颜料亦在所非宜，封建社会的民居尤不能与皇家同一享受，因此色彩只好以雅静为归，用清幽胜丽，设计上用少胜多的办法了。此种色彩其佳处是与整个园林轻巧的外观，灰白的江南天色，秀茂的花木，玲珑的山石，柔媚的流水，都能相配合调和，予人的感觉是淡雅幽静。这又是江南园林的特征了。

中国园林还有一个特色，就是设计者考虑到不论风雨明晦，景色咸宜，在各种自然条件下，都能予人们以最大最舒适的美感。除山水外，而楼横堂列，廊庑回缭，阑楯周接，木映花承，是起了最大的作用的，给人们提供了在各种自然条件下来欣赏园林的条件。诗人画家在各种不同的境界中，产生了各种不同的体会，如夏日的蕉廊，冬日的梅影、雪月，春日的繁花、丽日，秋日的红蓼、芦塘，虽四时之景不同，而景物无不适人。至于松风听涛，菰蒲闻雨，月移花影，雾失楼台，斯景又宜其览者自得之。这种效果的产生，主要是设计者对文学艺术的高度修养而与实际的建筑以相结合，使理想中的境界付之于实现，并撷其最佳者渲染扩大，因此叠石构屋，凿水穿泉，栽花种竹，都是向这个目标前进的。文学家艺术家对自然美的欣赏，不仅是一个春日的艳阳天气，他们对任何一个季节时间，都要将它变成美的境地，因此在花影考虑到粉墙，听风考虑到松，听雨考虑到荷叶，月色考虑到柳梢，斜阳考虑到漏窗，岁寒考虑到梅竹等，都希望理想中的幻景而付诸实现，故其安排一石一木，都寄托了丰富的情感，宜乎处处有情，面面生意，含蓄有曲折，余味不尽了。此又为中国园林的特征。

五

以上所述，系就个人所见，掇拾一二，提供大家参考。我相信，苏州园林不但在中国造园史上有其重要与光辉的一页，同时今日尚为广大人民作游憩之所。我们如何继承与发挥优良的文化传统，此份资料似有提出的必要。当然管见所及，定多不妥，还希望大家提出予以指正。

一九五六年十月，陈从周写竟初稿于同济大学建筑系建筑历史教研组

Analysis of Suzhou Gardens

◆

苏州园林初步分析
（英文）

Analysis of Suzhou Gardens

I

The origin of Chinese gardens may be traced to You（囿）and Yuan（园）of ancient China, or Yuan（苑）in the book *A Study of the Han Dynasty Customs* （汉制考）. All these Chinese characters mean "fenced land". According to the first Chinese etymological dictionary *Explanation of Characters*（说文，～ 2nd century AD）You（囿）and Yuan（苑）are designed for the purpose of keeping and breeding animals, while Yuan（园）and Pu（圃）are created as a means of growing fruits and vegetables respectively. People are assigned in charge of these lands. According to *Rites of Zhou*（周礼，～ 2nd century, BC）, the Primary Administrator（大宰）of the Heaven Office（天官）ordained nine duties to the commoners, the second of which is to manage Yuan（园）or Pu（圃）by way of cultivating trees and grass. Also recorded in *Rites of Zhou* is that the Manager of You（囿人）in the Earth Office（地官）takes charge of shepherding and breeding animals, whereas the Master of Load（载师）cares for transforming the land into Yuan（园）and Pu（圃）.

Construction of You（囿）or Pu（圃）started as early as the time of Xi Wei（豨韦）or Huang Di（黄帝）respectively. Later, during the period of the three ancient Chinese dynasties - Hsia, Shang and Zhou（22nd century to 17th century BC）, Yuan（苑）and You（囿）were designed as hunting grounds. For example, King Wen of the Zhou dynasty, Ji Chang（周文王，姬昌）, built a You（囿）for the people to collect firewood and hunt for wild birds and rabbits.

Starting from the Qin and the Han dynasties, the function of gardens was gradually shifted into places for exclusive enjoyment of the emperor, with elaborately designed and decorated buildings. Some of them were very large according to historic records. The first emperor of a unified China, Ying Zheng of the Qin dynasty（秦始皇，嬴政）constructed the Palace Qin across both banks of the Wei River, covering an area of 300 Li（a distance unit used in ancient China, one Li is equal to 0.5 kilometer）. Emperor Wu of the Han dynasty, Liu Che（汉武帝，刘彻）built Shang Lin Yuan（上林苑）, Gan Quan Yuan（甘泉苑）, as well as Tai Ye Chi（太液池）, to north of the Jian Zhang Palace（建章宫）.

The King Xiao of Liang, Liu Wu（梁孝王，刘武）started to put up rockery in his Tu Yuan（兔园）. The emperor Wen of the Wei dynasty, Cao Pi（魏文帝，曹丕）built Fang Lin Yuan（芳林园）and the emperor Yang of the Sui dynasty, Yang Guang（隋炀帝，杨广）constructed Xi Yuan（西苑）. The Emperor Yi Zong of the Tang dynasty, Li Cui（唐懿宗，李漼）set out to transform Yuan（苑）into a prototype of Chinese gardens by planting arbors and piling rockery in it. The garden Gen Yue（艮岳）built by the Emperor Hui Zhong of Northern Song dynasty, Zhao Jie（宋徽宗，赵佶）, is the most documented garden in historical texts.

After the Song dynasty migrated south, there were built various gardens such as Yu Jin（玉津）, Ju Jing（聚景）, Ji Fang（集芳）, in the area of Lin'an（临安，now Hangzhou 杭州）. The emperor Kublai Khan of the Yuan dynasty（元始祖，忽必烈）built Tai Ye Chi（太液池）of Long Live Hill based on the island Qiong Hua（琼华岛）of the Liao and the Jing dynasties. During the periods of the Ming and the Qing dynasties, in light of traditional customs, many new style gardens were brought into being for the sake of emperors, such as West Yuan（西苑）, South Yuan（南苑）, as well as Chang Chun（畅春）, Qing Yi（清漪）, and Yuan Min（Old Summer Palace, 圆明）in the western suburbs of Beijing.

The development of private gardens can be traced back to Yuan Guanghan（袁广汉）of the Han dynasty, who constructed a garden at the foot of Mount Bei Mang（北邙山）located north of Luoyang. This garden covered an area of four Li in the east-west direction, and five Li along the north-south direction. It was a huge garden with ranges of rockeries and a variety of animals kept within their bounds. Liang Ji（梁冀）built several Yuans and Yous which were distributed in an area thousands of li in extension, bordered by Hong Nong（弘农）to the west, Ying Yang（荥阳）to the east, Lu Yang（鲁阳）to the south, and He Qi（河淇）to the north. Treasury Minister of the Han dynasty, Zhang Luen（汉司农，张伦）, built the garden Jing Yang Shan（景阳山）, in close imitation of natural landscape, highlighting the achievements of the art of garden architecture. In addition, Ru Hao（茹皓）, an artisan of Wu region, built towers and pavilions around the grotesque rocks collected from Bei Mang（北邙）and Nan Shan（南山）, and channeled natural springs to water the garden flowers. The above-mentioned gardens were all initial attempts at imitating the sceneries of nature.

During the period of the Wei, the Jing and the Six dynasties, China was undergoing a dramatic change in ideology and was inflicted with the most incessant warfare in Chinese history. A class of scholar-officials, who maintained an elegant living style, enjoyed a simple and refined life and upheld Buddhism prevalent at the time, started to think of quitting politics and to spend time worshiping Buddha for self cultivation. They hankered after an idyllic life in nature while actually residing within cities. This philosophy led to a boom in poetry eulogizing nature and an emergence of paintings depicting wild mountains and rivers. Consequently, these officials started to set up gardens adjacent to their residence in order to pursue their ideal.

The preface to the essay *Back to Nature*（思归引序）by Shi Chong（石崇）reveals a couple of guiding principles in the design of his Jin Gu Yuan（Garden of Golden Valley, 金谷园）, i.e. to "avoid the hustle and bustle" and "exult in the beauty of nature". These thoughts also found their way into many other essays such as *Biography of Xiao Tong*（梁书·萧统传）in the Liang dynasty, *A Warming Note from Xu Mian to Zi Song*（徐勉戒子崧书）in the South Liang dynasty, and *Rhapsody of Garden by Yu Xin*（庾信小园赋）of the Liang dynasty. The gardens built during the Tang dynasty, such as Lang Tian Villa（蓝田别墅）of Song Zhiwen（宋之问）, Ping Quan Villa（平泉别墅）of Li Deyu（李德裕）, and Wang Chuan Villa（辋川别业）of Wang Wei（王维）, all pride in a magnificent view of bamboo islets and flowery docks and display an interest in verdant creeks and clear water. In these gardens, the man-made scenes resemble strongly those wrought by the forces of nature. On the other hand, in his Cao Tang（草堂）, the poet Bai Juyi（白居易）incorporated elements of nature into the design by borrowing scenery from the surrounding. The article *Famous Gardens of Luoyang*（洛阳名园记）written by Li Gefei（李格非）of the Song dynasty records major gardens in Luoyang region from the Sui and the Tang dynasties to the Northern Song dynasty, such as the Garden of Duke Fu Zheng（富郑公园）. Another article of the same period by Zhou Mi（周密）, entitled *Records of Gardens in Wuxing Area*（吴兴园林记）documents the gardens in the Southern Song dynasty in Wuxing area, such as the Garden of Minister Shen（沈尚书园）. The garden design as chronicled in these articles is already very close to that of the present day.

There was an explosive growth of gardens during the Ming and the Qing dynasties. These gardens are too numerous to be cited individually, but it is worthwhile to mention a few here just as examples, such as Shao Yuan（勺园）and Man Yuan（漫园）in Beijing; Ying Yuan（影园）, Jiu Feng Yuan（九峰园）and Ma Shi Ling Long Guan（马氏玲珑馆）in Yangzhou; An Lan Yuan（安澜园）in Haining; Xiao You Tian Yuan（小有天园）in Hangzhou; as well as Dong Yuan（东园）and others described in the essay *Tour of Various Gardens in Jingling*（游金陵诸园记）written by Wang Shizhen（王世贞）of the Ming dynasty.

Quite a few ancient gardens have survived up till now, such as Gao Yuan（皋园）in Hangzhou, Shi Yuan（适园）, Yi Yuan（宜园）, Xiao Lian Zhuang（小莲庄）in Nanxun, Yu Garden（豫园）in Shanghai, Yan Yuan（燕园）in Changshou, Gu Yi Garden（古猗园）in Nanxiang and Ji Chang Yuan（寄畅园）in Wuxi. Most of them are found in Suzhou and its surrounding areas. In the following we will focus on the historical development of Suzhou gardens for illustration.

II

The development of Suzhou as a political, economic, and cultural center started during the reign of the Wu State in the Spring and Autumn Period of ancient China. It continued to grow through both the Han and Jing dynasties. The Parasol Garden built during the reign of the Wu state in the Spring and Autumn Period and the Garden of Gu Pijiang（顾辟疆）constructed in the Jing dynasty are the two precursors of early Suzhou gardens. The region south of Yantze grew especially prosperous in the Six Dynasties era. The three major cities in this region are Yangzhou, Nanjing and Suzhou. In due course each of the three cities developed its economy in its own way: the one as a center of commerce, the other as a base for silk textile and the artisan handcrafts, and the still another as a consumer town for bureaucrats. Suzhou became a site for handicraft manufacture and for consumer service for officials and land owners.

During the Sui dynasty, Emperor Yang, Yang Guan（隋炀帝，杨广）constructed the Grand Canal, thus improving the transportation of goods between the north and the south China. Since the Tang dynasty the maritime trade increased significantly, making the region south of Yangtze more prosperous than the time of the Six Dynasties. Toward the end of the Tang dynasty, while the central provinces of China were ravaged by constant warfare, the area of Wu and Yue states as well as the Southern Tang was less affected and remained stable and thriving. The status quo continued through the Northern Song dynasty. When Emperor Gao of the Song dynasty, Zhao Gou（宋高宗，赵构）, fled south, Suzhou became the seat of Ping Jiang Prefecture（平江府）. This city also served as the residence of Emperor Zhao Gou during this period. Moreover, there was a growing trend of land ownership consolidation at the time and, as a result, there appeared numerous mansions of rich families. All these conditions favored the expansion of Suzhou gardens at the time. During the period of both the North and South Song dynasties there emerged a number of famous gardens built in the Suzhou region. Su Sunqing（苏舜钦）constructed the garden Can Lang Ting（沧浪亭, The Canglang Pavilion）on the site of Qian family's garden of the Wu-Yue period. On another site of the same Qian family's garden, Zhu Changwen（朱长文）built Le Pu（乐圃）. Mei Xuanyi（梅宣义）built Wu Mu Yuan（五亩园）. Zhu Mian（朱勔）not only built and operated Gen Yue Garden（艮岳）for Zhao Jie（赵佶）, but also constructed Tong Le Yuan（同乐园）for his own enjoyment. Wang Huan（王焕）, Governor of Ping Jiang Prefecture, erected pavilions and dug ponds as a garden annex in the northern part of the government building.

During the Yuan dynasty, the provinces of Jiangsu and Zhejiang continued to be a hub of wealth and prosperity. The activity in garden design and construction still ran brisk. The garden Shi Zi Lin（狮子林, The Lion Forest Garden）was one in case. During the Ming and the Qing dynasties land ownership consolidation became even stronger. Not only was Suzhou a major manufacturing site for silk textiles and a variety of artisan handcrafts since the period of the Tang and the Song dynasties, but also it had a high concentration of land owners and bureaucrats. From this region came a significant portion of people who entered into the service of civil servants（bureaucrats）through the Ke Ju（科举, imperial exam) system. This city led the country in the number of Zhuang Yuan (状元, highest honor from imperial exam) during the Qing dynasty. Upon retirement from the civil service, these civil servants purchased land and built residences. Profiting from land ownership as well as business operation, these people had gardens widely constructed for their enjoyment of leisure. They were also a major body of consumers of the handcrafts produced in this region. This general market condition is similar to that of Luoyang of the Sui and the Tang dynasties, Wuxing of the Southern Song dynasty, and Nanjing of the Ming dynasty.

Besides, Suzhou is endowed with the beauty of natural environment. It is situated in a region crisscrossed by waterways and dotted with numerous lakes. It is easy to find water sources and natural springs. In addition to the abundance of fresh water, the soil in this area is extremely fertile and conducive to the growth of flowers and plants. There are a number of quarries in the region providing easy access to masonry material. The famous Hu Shi (湖石, lake stone) can be found at Rao Feng Mountain (尧峰山), East Dong Ting (洞庭东山) and West Dong Ting (洞庭西山) mountains. There are also a number of accessible quarries located in neighborhoods further away from Suzhou. But the quality of stones produced there is not as fine as that found in the quarries near the city. These quarries included Huang Mountain of Jiangyin (江阴黄山), Zhang Gong Cave of Yixin (宜兴张公洞), Chui Mountain (镇江圌山) and Da Xian Mountain (镇江大岘山) of Zhengjian, Long Tan Lake at Jurong (句容龙潭), Qing Long Mountain of Nanjing (南京青龙山), and Ma An Mountain of Kunshan (昆山马鞍山). The priority choice of masonry for garden construction in Suzhou is lake stones for the reason of their extremely picturesque shapes. *The biography of Dai Yong* (戴颙) in *Book of Song* (宋书) notes "Yong came to live in the Wu region. The local scholars laid their heads together to design a house for him. They collected stones from around, channeled mountain springs to water trees planted and dug a pond after a pond. Before long, as the plants flourished, the scene of the villa became a part and parcel of the natural milieu."

In addition, Suzhou is a major center of art and scholarship in the nation. It prides itself on having an abundance of poets, writers, and painters. The scholars not only created their own ideas, but also drew wisdom from their residence advisors known as Men Ke（门客）. It is recorded in the book *Record of Wu Region Practices*（吴风录）that: "The descendants of Zhu Mian（朱勔）live on the hill of Hu Qiu（虎丘）. They chose the occupation of stone craftsmanship in garden construction for the high officials and royalties. The locals call them Hua Yuan Zhi（landscapers, 花园子）." Another passage from *Miscellany in the Gui Xing Year*（癸辛杂识）by Zhou Mi（周密）states: "The workers from Wuxing region are called rockery masons. This might be the legacy of Zhu Mian（朱勔）." These records demonstrate the prevalence of both architects and artisans in the region. The most prominent designers since the Song dynasty were almost all from the provinces of Jiangsu and Zhejiang. Those designers worth mentioning are Yu Cheng（俞澂）, Lu Dieshan（陆叠山）, Ji Cheng（计成）, Wen Zhenghen（文震亨）, Zhang Lian（张涟）, Zhang Ran（张然）, Ye Yao（叶洮）, Li Yu（李渔）, Chou Haoshi（仇好石）, and Guo Yuliang（戈裕良）. Even today the majority of artisans specializing in rockery design come from the cities of Nanjing, Suzhou and Jinghua. Due to its long heritage in the Suzhou region, the rockery designers from Suzhou are the best known in Jiangsu and Zhejiang areas. According to the writings in the *Record of Wu Region Practices* the art of landscaping is not just limited to the scholars who have the wealth to design and construct the garden villas, it also finds its way among "even the commoners from Lu Yan（间阎, countryside）who still decorate their homes with small rocks". This shows the deep love of the local people for nature.

The city of Suzhou has inherited the largest and well-preserved collection of gardens within its city limits. It may well be called the city of gardens if we are able to make an inventory of the gardens still prevalent there. The gardens of Suzhou are the best amongst present day Chinese gardens. It is no wonder the gardens of Suzhou are often celebrated with the words "The gardens in the South of the Yangtze are the best in China, and the gardens of Suzhou are the best of South of the Yangtze ". Over a period of five years I was able to conduct a series of survey and research on these gardens. In addition I was also involved in the restoration of several gardens. In the last two summers I came here with the students of the Department of Architecture at Tongji University to conduct onsite teaching and surveys focusing on ancient architecture and gardens. I availed myself of the long field work gathering written and photographic data for reference. The gardens Zhuo Zheng Yuan (The Humble Administrator's Garden, 拙政园) and Liu Yuan (The Lingering Garden, 留园) are used to illustrate prototypical design practices and evolution of Suzhou gardens. Other smaller gardens will be referred to wherever necessary. The verses used to describe the photos are collected or adapted from poems of the Song dynasty period. They are embellishments to the garden sceneries.

III

Zhuo Zheng Yuan (The Humble Administrator's Garden, 拙政园)

The Humble Administrator's Garden is located on Donbei Jie (东北街, North East Street), between the two city gates, Lou and Qi (娄齐二门), in the northeast corner of Suzhou city. It was constructed on the remains of Da Hong Temple (大宏寺) by Wang Xiancheng (王献臣) who acquired the land during Jia Jing (1522 AD—1566 AD, 嘉靖) period of the Ming dynasty. The name "Humble Administrator" (拙政) is derived from Pan Yue's (潘岳) phrase "A humble way being an administrator" (拙者之为政).

This garden was sold to Xu family of Lizhong (里中徐氏) by Wang Xiancheng's son in order to settle his gambling debt. Its ownership then fell into the hands of Chen Zhiling (陈之遴) from Haining (海宁) in the early part of the Qing dynasty. The garden became the official residence for the garrison general after Chen Zhiling was banished to a faraway outpost for penal servitude. After that, it turned into the administrative depot of military supply for local circuit.

This garden was for a time the residence of Wang Yongnin (王永宁), the son-in-law of Wu Sangui (吴三桂). It was repossessed by the government after Wu Sangui's rebellion against the emperor. During the reign of Emperor Kang Xi (康熙), the garden was used as the official residence of the administrator of the circuit of Suzhou, Songjiang, and Changzhou. Emperor Kang Xi visited this garden during his southern tours of the empire.

The garden became divided into many small residences when the position of administrator for circuit of Suzhou, Songjiang, and Changzhou was later eliminated. Jiang Qi (蒋棨) brought the various parts of the garden together in early years of Emperor Qian Long's (乾隆) reign and had it renamed as Fu Yuan (复园, Recovered Garden). It was then sold to Cha Shitan from Haining (海宁查世倓) during the reign of Emperor Jia Qing (嘉庆). Later this garden changed hands again and became the property of Wu Songpu from Pinghu (平湖吴菘圃).

After Tai Ping Tian Guo (太平天国, Heavenly Kingdom of Great Peace) took control of Suzhou, the garden became a part of Prince Zhong's residence (忠王府). The garden became the property of the imperial government after the rebellion of the Tai Ping Tian Guo. It served as the venue of Ba Qi Clan Association (八旗奉直会馆) starting in the tenth year of Tong Zhi (同治, 1871 AD). The garden was restored to its original name, the Humble Administrator's Garden. During this period the west section of the garden became the property of Zhang Luqiang (张履谦) and was named Bu Yuan (Addition Garden, 补园). The two gardens have since been reunited into one after 1949.

The highlight of the Humble Administrator's Garden is water. Over sixty percent of the garden is covered by water. As a result of vast water coverage, eighty to ninety percent of the main buildings in the garden are built along water front (Page 275, Site Plan of the Garden). This is documented in the article *A Record of the Humble Administrator's Garden* (拙政园记) by Wen Zhengming (文徵明) in this way: "There were many marsh lands nestling in the northeastern section of the city. This is especially true for the area between the city gates Lou and Qi. Water accumulates in these areas to form natural ponds. These ponds were channeled together with little effort and lined by trees around them. ⋯" It can be seen from these lines that the garden was designed based on the then existing topography.

The practice of incorporating existing topography into the design is also described by Ji Cheng (计成) in his book *Yuan Ye* (*Craft of Gardens*, 园冶) written during the late Ming dynasty. It is said in Xiang Di section (Site Locating and Evaluation, 相地) of *Yuan Ye* that "Pavilions and platforms should be located on high grounds, and ponds should be dug at low lands." The choice of the site for the Humble Administrator's Garden on or around water is well justified because of the readily available low lands in the area.

Incorporating existing topography into garden site planning is a common practice found in many gardens other than the Humble Administrator's Garden in Suzhou. The garden Wang Shi Yuan（网师园, The Master-of-Nets Garden）at Kuo Jie Tou Xiang（阔阶头巷）has eighty percent water coverage. The same is true of Wu Mu Yuan（五亩园）at Ping Gate（平门）, in which several small ponds were linked together to form a vast water surface. All of these gardens are invariably created by taking advantage of existing topography.

Looking for underground spring is another good way for acquiring a fresh water source. In the garden Huan Xiu Shan Zhuang（The Mountain Villa with Embracing Beauty, 环秀山庄）on Jing De Rd（景德路）the water source is an underground natural spring found when digging the pond. This was done by the owner Jiang Ji（蒋楫）during the reign of Emperor Qian Long（乾隆）of the Qing dynasty. This spring was named "Flying Snow"（飞雪）by Jiang Ji.

The Humble Administrator's Garden is made of three distinct parts. Its center section is the main part of the garden. This section contains the majority of the original designs that have survived over the years. The west section is the portion of garden that used to be Bu Yuan（Addition Garden, 补园）of Zhang family. This section has received many major modifications but maintained its original style. These changes do not diminish the glamour of a well designed garden. The east section was Gui Tian Yuan Ju（Retreat House, 归田园居）of Wang Xingyi's（王心一）during the Ming dynasty, which has long been neglected, and is now being restored.

Yuan Xiang Tang（the Hall of Distant Fragrance, 远香堂）（Page 86）is the principal building in the center section of the garden, with a four-sided three-bay–wide hall, and a single eave gable-on-hip roof（单檐歇山）. Looking south from the hall one enjoys a view of a small pond with rockery. There is a lush green shade provided by several clusters of magnolia. An old elm tree stands by the rock nearby an undulating wall（云墙）. A line of dark green bamboo hugs the rockery. The slender winding veranda follows the curve of the hill rocks. It leads the visitor into the garden. This type of site planning not only creates an impression of entering a natural forest（Page 84, upper）for the visitors, but also effectively hides the garden entrance for people looking south from the main hall by using the rockery as a natural screen. This design technique is similar to the use of Zhao Bi (Screen Wall, 照壁) or Ping Feng (screen, 屏风) at the entrance of a residence.

Sitting inside the Hall of Distant Fragrance and looking at north, one can get a glimpse of major sceneries of the whole center section of the garden, with Xiu Qi Ting (Paeonia Suffruticosa Pavilion, 绣绮亭, Page 90) to the east, Yi Yu Xuan (Bamboo Pavilion, 倚玉轩, Page 86) on the west side, a large lotus pond bordering the north side, as well as two pavilions atop islets across the pond, named Xue Xiang Yun Wei Ting (Snow-Like Fragrant Prunus Mume Pavilion, 雪香云蔚亭, Page 105) and Dai Shuang Ting (Orange Pavilion, 待霜亭, Page 104). Along the north-to-south axis of the Hall of Distant Fragrance the garden's profile goes up and down with accentuating motifs. In particular, the islets in the pond are surrounded with flowing water, veiled by dwarf bamboo, and adorned by lime stones along the jagged shore. The whole view stretches one's boundless imagination as if one had found oneself in a true natural scene of lake.

Strolling along the waterside trail eastward from the Hall of Distant Fragrance takes one to the bridge Yi Hong Qiao (Leaning against Rainbow Bridge, 倚虹桥, Page 96). This bridge is a part of the original Ming dynasty construction. It is characterized by the low balustrade and close proximity of the bridge surface to water surface. After crossing the bridge one arrives at the pavilion Yi Hong Ting (Leaning against Rainbow Pavilion, 倚虹亭). This pavilion is built with its east side against a wall and the other three sides to open terrain. Therefore, it is also called East Half Pavilion (东半亭).

The passage northward from the Leaning against Rainbow Pavilion leads to the Pavilion Wu Zhu You Ju (Secluded Pavilion of Firmiana Simplex and Bamboo 梧竹幽居, Page 102). This is a pavilion with a pyramid-shaped roof and a ring-shaped gate on each of its four sides. Here is the east end of the center section of the garden. Looking at the west of the garden through the two ring-shaped gates one gets an extraordinary view of the beauty around the pool (Page 99), with a dimly visible sight of the West Half Pavilion (西半亭) located far away on the opposite side of the garden across the pond. The West Half

Pavilion is well known as Bie You Dong Tian (Another Cave World, 别有洞天) which refers to the moon gate in this Pavilion since this moon gate leads into the west section of the garden.

At the Secluded Pavilion of Firmiana Simplex and Bamboo, one can see the reflection of the gate "Another Cave World" in the water, and the Pagoda Bei Si Ta (North Temple Pagoda, 北寺塔) rising sheer into the sky behind the West Half Pavilion. This is a spectacular use of the scene borrowing (借景) technique. Besides there is a sharp contrast presented to the viewer at this vantage point (Page 97). The left side of the pool unfolds a group of elaborate buildings including the Hall of Distant Fragrance, the Bamboo Pavilion, and Xiang Zhou (Fragrant Isle, 香洲). On the right side lie two small islets in a totally natural milieu. "This is the best use of contrast in a garden design." as noted by Master Liu Dunzhen (刘敦桢). As a matter of fact, the physical distance between the east and the west sides of the pond is not particularly large. But the visual effect of the design provides an illusion of depth. This is partly due to the construction of a winding stone clapper bridge across the pond. One's sight lingers on the bridge before reaching the west shore. On the other hand, the strip of water between the north and the south shores is slender, and the magnificent buildings on the south contrast wonderfully with the natural verdant hills on the north. All these give people a profound feeling of depth and extension. With aged elm standing by the shore, swaying willow leaning onto the water, a new cluster of pavilions and halls looming through an elegant gate in between, what catches one's eye is another wonderland on the opposite bank.

Walking along the passage westward from the Secluded Pavilion of Firmiana Simplex and Bamboo and crossing the Three Zigzag Bridge (三曲桥) one descends on a small triangle shaped islet. Taking the winding trail up hill one arrives at a hexagonal shaped pavilion nestling among dwarf bamboos at the peak, which is named Dai Shuang Ting (Orange Pavilion 待霜亭, Page 104). To its east and across the water is the Pavilion Lu Yi Ting (Green Ripple Pavilion, 绿漪亭, Page 100). On the west it is matched by Xue Xiang Yun Wei Ting (Snow-Like Fragrant Prunus Mume Pavilion, 雪香云蔚亭, Page 105), which is a rectangular shaped pavilion located on an elliptical shaped islet. A small bridge across a small creek connects the two islets (Page 108). Enchanted by aged trees leaning in a bamboo grove, clear stream gurgling and orioles twittering in the arbor brush, one feels a sudden isolation from the mundane world.

The trail winding down from the Snow-Like Fragrant Prunus Mume Pavilion leads to the Pavilion He Feng Si Mian Ting (Pavilion in Lotus Breezes, 荷风四面亭, Page 125). This is another hexagonal shaped pavilion. It is at the confluence point of three passages, with connecting zigzag bridges on two of the passages. The way forward heads for the Bamboo Pavilion, while the way backward brings one to Jian Shan Lou (Mountain-In-View Tower, 见山楼) and "Another Cave World" through the winding veranda Liu Yin Lu Qu (A winding way with willow shades, 柳阴路曲, Page 116). The Mountain-In-View Tower (Page 109) is a two-storey building with a double eave gable-on-hip roof (重檐歇山顶). Its second storey is reached via a staircase formed by rockery on the exterior. This building is located on the northwest corner of the center section of the garden. It is well separated from other structures in all directions, providing a wide-angled view from the top of the building. The whole city comes into focus when looking northward.

Due to the large size and expansive water surface of the Humble Administrator's Garden, such a two-storey tower built in it does not look overly tall. What's more, its reflection in the limpid water is a welcome addition to the scenery. However, in designing a two-storey building in a garden due attention should be paid to its architrave arrangement at elevation. One should adopt more horizontal lines running in parallel to the water surface in order to achieve a better consistency.

This type of vantage viewpoints can be found in many Chinese gardens. For example the pentagonal shaped pavilion built in Ban Yuan (Half Garden, 半园) at Zhong You Ji Xiang (中由吉巷), and Shan Mian Ting (Fan Pavilion, 扇面亭, Page 206) found in Shi Zi Lin (Lion Forest Garden, 狮子林) are both located on high peaks in the corner of the gardens.

The Fragrant Isle is traditionally known as a land boat, a building formed in the shape of a boat though it cannot float as a boat on water (Page 117 and 118). There is a large mirror lodged in the bay of the building. When viewed from the Bamboo Pavilion, one cannot tell whether one sees a true subject or a reflection in the mirror. It also gives a sense of depth to a relatively shallow area. The upper part of the two-storey structure of Fragrant Isle, which is called Cheng Guan Lou (澄观楼, clear water viewing tower), is in an advantageous position for enjoying faraway scenery.

To the south of the Fragrant Isle is the Pavilion De Zhen Ting (True Nature Pavilion, 得真亭, Page 123), in which there is also a mirror that creates similar effect of illusion. In this area there is a narrow strip of water. The stone bridge Xiao Fei Hong (Small Flying Rainbow Bridge, 小飞虹, Page 120) crosses over the water surface and breaks it into two parts. To the south of the bridge, a three bay pavilion Xiao Cang Lang (Small Cang Lang, 小沧浪) perches over the water and further divides the water surface. With open spaces beneath the bridge and pavilion, one does not feel any constraint in this narrow area. Instead, the structural arrangement gives people a free and unrestrained feeling.

Looking from the Small Cang Lang and through the Small Flying Rainbow Bridge and the Pavilion Yi Ting Qiu Yue Xiao Song Feng Ting (Autumn Moon and Windy Pine Pavilion, 一庭秋月啸松风亭, Page 120 and Pages 122), one has a panoramic view of a large expanse of water with the reflection of the Pavilion in Lotus Breeze in it, the profile of the Fragrant Isle as well as the Mountain-In-View Tower afar (Page 121). This multiple level scenery gives a sense of looking at a grand view through a small portal, and cheers up the viewer with its clear open space.

Pi Pa Yuan (Loquat Garden Court, 枇杷园) lies on the southeast side of the Hall of Distant Fragrance. It is separated by an undulating wall from the rest of the garden. There is a moon gate connecting the two sections (Page 87). Jia Shi Ting (Loquat Pavilion, 嘉实亭) and Ling Long Guan (Hall of Elegance, 玲珑馆) are both situated in the front portion of the Court. Sitting in the Courtyard and looking at the Snow-Like Fragrant Prunus Mume Pavilion through the moon gate one has the impression that the pavilion is embedded in a ring. This is among the best in paired scenery (对景, Page 88). It seems as if we were sitting within a painting, with the tall scholar tree and pavilions by the low wall fore-grounded indeed. This is an extremely good use of scene borrowing (借

景). The courtyard ground is paved with cobblestones in an elegant fashion. Unfortunately the rockery along the wall has lost its original shape during reconstruction; the upper part of the undulating wall lacks the traditional ending, and the turn in the wall lacks finesse (Page 87).

A winding veranda on the side of the Hall of Elegance leads to Hai Tang Chun Wu (Malus Spectabilis Garden Court, 海棠春坞, Page 92). This is only a two bay wide building (Page 93). Standing by the steps are an aged tree, a crabapple, and some scattered pieces of rockery; surrounding the Court is a veranda; veiled behind the tracery windows are pavilions and bowers, water and rock (Page 94-95). With such careful arrangements in a closed and confined space, the Court looks delicate and elegant at every point, and spacious and smooth in every direction. This demonstrates the ingenuity of Chinese garden design, which enables sparsity artistically spaced with density and *multum* in *parvo*.

A wall was built between the west and center sections when the Humble Administrator's Garden was split to two gardens. Tracery windows were added in the wall after these two sections were united. Some minor modifications had been made in the west section when it was split out, in order to create an independent entity. A winding verandah over the water was built along the wall. The veranda rises and falls while snaking along the water, making people feel like treading the water waves while walking through the veranda (Page 132). This is the best touring veranda design amongst all found in Suzhou gardens.

The San Shi Liu Yuan Yang Guan (Hall of Thirty-six Pairs of Mandarin Ducks, 三十六鸳鸯馆) and Shi Ba Man Tuo Luo Hua Guan (Hall of Eighteen Camellias 十八曼陀罗花馆) is a twin hall (Page 141). This is the main structure in the west section, three bays wide, with a gable-on-hip roof at exterior. But there is no roof ridge seen in the interior. Instead, four fold arc ceilings (卷棚四卷, Page 317) are used. Additionally, there are four rooms with individual pyramid roofs at the four corners of the building. As a result its design is unique in the country.

This twin hall was built when the west section was split out, out of the need of a main structure in a stand-alone garden. In order to fit such a big hall into a constrained space, it was necessary to downsize the front portion of the hall and extend the rear of the hall over the water. Consequently, the water surface is reduced. The short distance to the hills across the pond and the limited clearing around the building makes the hall inconsistent with the rest of the garden to a certain degree. These are the defects in an otherwise perfect design.

This hall was used by the owner to dine and entertain guests. The special interior roof structure enhances its acoustic effects. The rooms at four corners provide not only a way to reduce the wind rush while people entering and leaving the hall, but also a space for the servants in waiting on call, or for the performers as a temporary backstage. The designer has taken all things into consideration when designing the structure. Its interior decorations along with those found in another building, Liu Ting Ge (Stay and Listen Pavilion, 留听阁) are among the most ornate found in Suzhou. The Hall of Eighteen Camellias is an ideal spot for enjoying mandragora flowers during early spring. And in the Hall of Thirty-six Pairs of Mandarin Ducks watching mandarin ducks playing amongst the lotus is a favorite pastime during summer. It is fitting to have these two halls with one facing the south and the other the north.

The Fu Cui Ge (Floating Green Tower, 浮翠阁) sits aside the water from the twin hall. It is a two storey octagonal structure. A bird's eye view of the garden is provided from the top of the building. It is unfortunate the structure is un-proportionately tall and does not fit in scale with the rest of the garden. Down from the Floating Green Tower and across a creek, two small pavilions are crouching on a small hill. They are Li Ting (Indus Calamus Pavilion, 笠亭, Page 144) and Shan Ting (Fan Pavilion, 扇面亭, Page 145). Fan Pavilion is also called "With Whom Shall I Sit?" Pavilion (与谁同坐轩). Being located at a bend of a slender stream, this Pavilion fits for leisure and relaxation.

Walking northward along the winding verandah over the water one reaches Dao Ying Lou (Tower of Reflection, 倒影楼, Page 135). This is a two storey building with a gable-on-hip roof. Sitting opposite to its south is a hexagonal pyramid-shape roofed pavilion, Yi Liang Ting (Good for Both Families Pavilion, 宜两亭, Page 137). The beautiful reflections of both buildings in the water form a perfect scene pairing.

Another scene pairing of similar type can be found to the west of the twin hall. In this case, on the south is the pavilion Ta Ying Ting (Pagoda Reflection Pavilion, 塔影亭, Page 149) while on the north is the building Liu Ting Ge (Stay and Listen Pavilion, 留听阁, Page 148). Both sets of building and pavilion pairs are astride a narrow body of water. The exterior shapes of the two pairs are extremely similar although the latter pair is located on lower ground than the former.

When the west section was split off from the garden, a new entrance was opened at the south side of the Pagoda Reflection Pavilion, which serves also as a passage to Zhang residence.

The east side of the garden has long since fallen to disrepair. Its rebuilding is being planned, and will not be discussed here. (Note by Translator: The east section has been rebuilt and opened to public. Indeed, the entrance to the garden is now in the east section.)

Liu Yuan (Lingering Garden, 留园)

The Lingering Garden is located on Lingering Garden Road just outside of city gate Chang (阊门). It was originally known as East Garden of Xu Taishi (徐泰时) during the Ming dynasty. During the reign of Emperor Jia Qing (嘉庆) of the Qing dynasty (circa 1800AD) this garden was rebuilt by Liu Shu (刘恕). It was called Han Bi Shan Zhuang (寒碧山庄) at this time, but was commonly known as the Garden of Liu. There were twelve rockery peaks in this garden which were crafted from the best lake stones found in the Tai Hu Lake (太湖). This garden then became the possession of Sheng Kang (盛康) in 2nd year of Emperor Gaung Xu's reign (光绪, 1876 AD) and renamed as Lingering Garden. Occupying an area of fifty Mu, it is the largest garden in Suzhou (Page 325).

This garden is divided into four parts, i.e. center, east, west, and north in light of their respective locations. The center section consists of a large pond surrounded by hills, rockeries, and buildings, which are stringed together by a series of verandas and bridges. The east section features its variety of buildings. In this part large and magnificent halls are accentuated by small studies and lined by rugged rockeries, resulting in a layout full of variation. The west section is mainly a man-made hill covered by a maple tree grove and dotted with a few pavilions. The south side of the hill is ringed by water to imitate Tao Yuan of Wu Ling of the Jing dynasty（晋人武陵桃源）. This section is partitioned from the center section by an undulating wall. The red leaves poking above the pale white walls look like red clouds sailing from distant sky. It forms the best borrowing scene（借景）in the center section. The structures currently found in the north part are all recent reconstruction. These buildings lack character in design. This section will not be discussed in detail.

Center Section: Entering the garden's front gate and passing through two small courtyards, one arrives at Lu Ying（Green Shade Pavilion, 绿荫, Page 155）. Through the tracery windows on the north side, one has a dim view of the hills and buildings in the garden. Going westward one reaches the main building of the center section, Han Bi Shan Fang（Hanbi（Be imbued with the green）Mountain Villa 涵碧山房, Page 157）. This is a three bays wide building, with a gabled roof with flushing purlins（硬山造）. On its south side is a small courtyard with a peony flower bed in the center while on its north side is a terrace bordering a lotus pond. Abutting against the Villa on the east is Ming Se Lou（Pellucid Tower, 明瑟楼, Page 159）. This is an ornate two storey building with a single sided gable-on-hip roof（单面歇山顶, Page 333, upper）. The second story can be reached via a scaling stair.

Continuing west from the Hanbi Mountain Villa and climbing up a veranda, one ascends to Wen Mu Xi Xiang Xuan（Osmanthus Fragrance Pavilion, 闻木樨香轩, Page 163）. This location affords a panoramic view of the center section and the surrounding areas. On the east side there is a spectacular array of buildings such as Qu Xi Lou（Winding Stream Tower, 曲谿楼）, Xi Lou（Western Tower, 西楼）, Qing Feng Chi Guan（Refreshing Breeze Pavilion by the Lake, 清风池馆）, Ji Gu De Geng Chu（Study of Enlightenment, 汲古得绠处）and Yuan Cui Ge（Distant Green Tower, 远翠阁）. These buildings, high or low and nestling among or snuggling against trees and rocks, enhance each other's beauty like a charm.

On the south side of the Osmanthus Fragrance Pavilion, pavilions and towers adjoin verandas and latticed window walls one after another along the water. The Pellucid Tower juts out to the water slightly, while the north terrace of the Hanbi Mountain Villa protrudes into the pond even further. This layered arrangement seems embosoming the water in the center and leaves a beautiful reflection of scene in the lucid pond.

The veranda north of the Osmanthus Fragrance Pavilion turns eastward at the north end of the garden and leads to the Distant Green Tower located in the northeast corner of the center section (Page 169). This tower serves the same purpose as the Mountain-In-View Tower in the Humble Administrator's Garden, though this building sits on land while the other by the water. Being located next to the east section of the garden and separated only by a wall adorned with latticed windows, the Distant Green Tower provides the east section with a magnificent borrowed scene (Page 168).

Standing in the middle of the water is Xiao Peng Lai (Small Fairy Isle, 小蓬莱), with Hao Pu Ting (Hao Pu Pavilion, 濠濮亭, Page 169) across the pond. Both of them are at the water level, which provides a significant contrast in height to the nearby hills and towers. The west wall of ground floor of the Winding Stream Tower is lined with brick-framed tracery windows (Page 170). Picturesque views through these widows keep varying while walking along the wall, creating fantastic and kaleidoscopic sceneries. In particular, when autumn comes, dark red maple leaves on the hill of the west section ebb and flow behind the undulating wall, making everyone thrilled.

Returning to Pellucid Tower and looking north, the hexagon shaped pavilion Ke Ting (Passable Pavilion, 可亭, Page 162) stands atop the hill across the lotus pond, with the long veranda leading towards the Distant Green Tower at the back (Page 166).

East Section: There are two major buildings in the east section. The first building is Wu Feng Xian Guan (Celestial Hall of Five Peaks, 五峰仙馆, or Hall of Nanmu, Page 173). This is a five bay wide hall with a gable roof with flushing purlins (硬山造). The interior decoration of this building is especially delicate and ornate (Page 174). It is one of the best old style halls in the region South-of-Yangtze. The courtyards on all four sides vary in size. There are rockeries in both the front and the back courtyards, which makes one feel as if he is facing cliffs when sitting in the hall. This design was not appreciated by Ji Chen of the Ming dynasty. He noted in the book *Yuan Ye* (*Craft of Gardens*, 园冶) that "every body wants to build rockeries in front of a hall. Having three high peaks stand and line up in a limited courtyard is funny indeed." In the Celestial Hall of Five Peaks, there are five rockeries in the front yard, making the court look overly crowded and devoid of lively interest. Under these conditions Ji Chen said also that one should "either put only small and exquisite rockeries amongst fine trees, or embed the rockeries within the wall, or plant some ivy or grass atop rockeries, in order to enhance perception of depth." This would be a more viable design practice in my own opinion. In the backyard, there is a limpid spring in front of the rockery, which provides an extremely quiet environment. It is unfortunate the spring has lost its source, and become intermittent. A veranda runs along the walls of this backyard. This veranda can lead to various directions and hence might disorient a person arriving at this point for the first time. The Distant Green Tower is clearly visible when looking left from this backyard. The yonder pavilion further demonstrates spaciousness of the garden (Page 168).

The building Ji Gu De Geng Chu (Study of Enlightenment, 汲古得绠处) is a one bay room adjacent to the Celestial Hall of Five Peaks. This structure has open void on all four walls, which provides a view of the sceneries in the center section. A flourishing umbrella shaped Chinese parasol tree stands erect in its courtyard adjoining the Celestial Hall of Five Peaks. This is a case of ingenious treatment of small space in design.

The buildings Huan Wo Do Shu Chu (The Return-to-Read Study, 还我读书处) and Yi Feng Xuan (The Worshipping Stone Pavilion, 揖峰轩) are both small courts on the east side of the Celestial Hall of Five Peaks. These two buildings serve as a transition in the space between the Celestial Hall of Five Peaks and Lin Quan Qi Shuo Zhi Guan (The Old Hermit Scholars' Hall, 林泉耆硕之馆). Here one can observe the full applications of small court design (Page 178-183). These small courts are surrounded by verandas and separated by hollow walls with brick framed windows. Inside the courts are arrayed fine trees accompanied with slender bamboos, as well as day lilies around slabs of rockery. All things here are well matched in size and in space, making the layout of the courts neat and tidy. One draws a lot of fun and enjoyment from these still objects and is often reluctant to depart.

The building east of the Worshipping Stone Pavilion is the second major building in the east section, the Old Hermit Scholars' Hall（林泉耆硕之馆）. The decoration in this building is extremely elaborate and rich（Page 184）. It is a five bay wide twin hall with a single eave gable-on-hip roof at exterior and two separate arc ceilings for each hall in the interior（Page 349 and 350）. The main hall faces north. Its main beam is made of flat wood decorated with carved patterns. In the south hall, the main beam is a round lumber without any carved decoration.

The north hall of the Old Hermit Scholars' Hall faces Huan Yun Zhao (The Cloud Bathing Pond, 浣云沼) as well as Guan Yun Ting (The Cloud-Capped Pavilion, 冠云亭) and Guan Yun Lou (The Cloud-Capped Tower, 冠云楼). Here stand the three most famous rockeries in Suzhou: Guan Yun Feng (The Cloud-Capped Peak, 冠云峰, Page 185), Xiu Yun Feng (The Mountainous Cloud Peak, 岫云峰, Page 190 Lower), and Rui Yun Feng (The Auspicious Cloud Peak, 瑞云峰, Page 191). These three peaks are a part of the original design of the Ming dynasty. They are the largest lake rockeries ever found in Suzhou.

The Cloud-Capped Pavilion is a hexagonal shaped pavilion located on the side of the Cloud-Capped Peak (Page 185 and 187). There are magnolia trees planted around the Pavilion. By climbing a scaling stair at north side of the Pavilion, one ascends to the Cloud-Capped Tower. At this place one enjoys an open view beyond the garden. The vast cultivated fields crisscrossed with footpaths and the dimly-visible pagoda image of distant Hu Qiu (The Tiger Hill), all come in sight at the window.

The buildings Zhu Yun An (The Standing Cloud Hut, 伫云庵) and Guan Yun Tai (Cloud-Capped Terrace, 冠云台) are located on the east and the west sides of the Cloud Bathing Pond, respectively. The Jia Qing Xi Yu Kuai Xue Zhi Ting (The Good-For-Farming Pavilion, 佳晴喜雨快雪之亭) is accessible from the Cloud-Capped Terrace through a moon gate. In the Good-For-Farming Pavilion, there is a six-piece Nanmu screen. The carving on the wood screen, rendered extremely intricate, is among the best decorations found in the Central Wu region (Page 192). Regrettably, the pavilion faces west, and so damages to this screen caused by weather and the sun cannot be avoided.

The southwest corner of the east section is an area of new creation, arrayed with some featureless rockeries, which cannot be compared with the structure of the original garden.

West Section: The west section looks like what remained of the East Garden constructed during the Ming dynasty. The hill is built by compacting earth and placing rocks at selected locations. This is just like the old technique summarized by Li Yu（李渔）as "small hill is built with rocks, while large hill with soil." As a result the large grove of maple trees growing on the hill receives sufficient nutrients from the earthen hill.

There are two pavilions situated within the maple grove. One is Shu Xiao Ting（The Free Roaring Pavilion, 舒啸亭, Page 193）which has a tapered cone shape roof. The other pavilion is the hexagonal shaped Zhi Le Ting（The Delightful Pavilion, 至乐亭, Page 194）. This pavilion is a transformation based on the Imperial Stela Pavilion（御碑亭）found at Fan's Ancestral Temple（范祠）at the Tian Ping Mountain（天平山）. The Free Roaring Pavilion crouches amongst the maple trees. The Delightful Pavilion snuggles on the northwest slope of the hill. The two pavilions are within sight of each other.

The south side of the hill is surrounded by a clear creek flowing in the shade of peach and willow trees growing along the banks. The original purpose of these arrangements is to create an out of the world environment. But it seems too artificial to be real here. It is not easy to make a man-made scenery to outshine a natural one, especially in a garden.

At the end of the creek is Huo Po Po Di（The Place of Liveliness, 活泼泼地, Page 197）. There is a pavilion built astride the water. The creek enters and submerges beneath the pavilion, which creates an illusion that the creek flows onwards. However, the rockeries in this area have lost most of their original outlook after various additions and repairs over the years.

North Section: The original structures in the north section have all been destroyed. And the new buildings in this section do not have any outstanding features.

IV

The typical gardens in the region south of Yangtze are small in scale, but are adorned with picturesque sceneries of steep cliffs and deep gorges along clear streams and emerald pools. Qian Yong（钱泳）said: "building a garden is just like writing a poem. It must be structured according to the rules." The best garden should have such architectural elements as hills, ponds, pavilions, halls, and verandas in a way that offers ever refreshing, enchanting and kaleidoscopic scenes and that does not fall into any pre-existing norms. Therefore, in terms of planning and space design, it is highly desirable to make the landscape look more expansive, eye-catching and ever-changing. In general, overall garden planning in Suzhou can be grouped into several categories as described below.

Water as the theme of the garden: In this type of design a water body is located in the middle of a garden surrounded by halls and pavilions connected by verandas. Small bridges sit astride the water to provide easy access. In a large pond, it is advisable to put up some islets. These types of designs are found in gardens such as Wang Shi Yuan（The Master-of-Nets Garden, 网师园）and the central part of Yi Yuan（The Garden of Pleasance, 怡园）. Chang Yuan at Miao Tang Xiang（The Garden of Free Will, 庙堂巷畅园）is situated within a relatively constricted space. Its design is as delicate as that of a bonsai. A small body of water lies in the middle, surrounded by verandas and buildings. The theme of the Lingering Garden seems to be the water body in its center section. However, Master Liu Dunzhen（刘敦桢）believes that, when this garden is viewed in its entirety, its theme is actually the group of structures in its east section. This opinion definitely deserves consideration.

Hills and rocks as the theme: This is an accepted practice when there is no readily available water source or low land within the garden. In such a case, it becomes a necessity to use hills and rocks as a theme to accentuate the landscape. The garden of Cheng family in Xi Bai Hua Xiang（西百花巷程氏园）is an example of such design. Huan Xiu Shang Zhuang（The Mountain Villa with Embracing Beauty, 环秀山庄）occupies a small land, where it is difficult to dig a large pond. As a consequence the designer picked hills and rocks as the theme and introduced springs to make them come to life. Well-nourished, the moss turns green and the grass becomes lush. The whole landscape is an image of nature.

In case of designing a garden with an expansive water surface, there is a lot of flexibility. Usually, this type of gardens is relatively large. Its site planning should not be restricted by any hard and fast rules. For the large gardens such as the Humble Administrator's Garden and Wu Mu Yuan (the Five Acres Garden, 五亩园) wide vision and variability are the keys to site planning. For example, the hills in the center section of the Humble Administrator's Garden are transformed with human hand into nature-like scenery with "flat hilltop and smooth slopes along the curved sand shores", as described by Zhang Lian (张涟) in his essay. Ji Cheng (计成) wrote in *Yuan Ye* (*Craft of Gardens*, 园冶) that "a garden should look as if created by nature though it is made through human efforts". Master Liu Dunzheng held that "water body in a garden should be concentrated, allowing only a small part diverted. In a small garden, concentration of water body is better than dispersed water system. In a large garden dispersing water body is viable but it should be done as a foil to the concentrated water area". Both the Humble Administrator's Garden and the Master-of-Nets Garden are exemplary designs of this type in Suzhou gardens.

Placement of a pond in front of Halls and with hills as background: In this deployment the halls are like waterside pavilions astride the waves. When sitting in the lounges, one enjoys a view of hills and rocks far across the pond. This type of layout is witnessed at Yi Pu in Wen Ya Long (The Garden of Cultivation, 文衙弄艺圃). The layout of Fang Chao Yuan (芳草园) on the east side of Bei Shi Ta (the North Temple Pagoda, 北寺塔) seems to have some resemblance to this type of design.

Placement of a pond and hills in the center surrounded by buildings and verandas: The array of buildings at different heights harmonize well with the hills and water in the center. This type of layout is rare in Suzhou. The only plausible example is the east side of Ou Yuan at Xiao Xin Qiao Xiang (The Couple's Garden Retreat, 小新桥巷耦园). The small garden of Han family on Dong Bei Jie (东北街韩氏小园) bears some resemblance to this layout. But there are buildings only on two sides. As regard to Ban Yuan at Zhong You Ji

Xiang（中由吉巷半园）and Song family garden at Xiu Xian Xiang（修仙巷宋氏园）, the buildings cluster only on one side.

The Suzhou gardens built in the Ming era closely resemble natural landscape. These design practices, as summarized by later architects such as Ji Cheng（计成）and Zhang Lian（张涟）, took advantage of the original scenery and made only minor modifications as needed. The rocks used to pile up hills are primarily Huan Shi（yellow rocks, 黄石）, as found in the two small hills as well as the mound under the Paeonia Suffruticosa Pavilion in the center section of the Humble Administrator's Garden. Although yellow rocks are not as exquisite as Hu Shi（lake stone, 湖石）in shape, the hills piled with yellow rocks look just as natural if done properly, without any affected air and any risk of collapse. Many fresh ideas and techniques may be discovered from studying these traditional designs in detail and hence can be applicable to today's garden construction.

During the Qing dynasty, the use of lake stones in rockery design became popular. So much so that there emerged a trend of coveting more rockeries and seeking the novelty of their structure. The number of rockeries and the quality of a single rockery thus became the sole criterion in evaluating the superiority of a garden. A typical example is the construction of the Garden of Pleasance, for which the best lake rocks from three garden ruins in Dong Ting Mountain area were used. However, in its northwest section, the stone skeletons of the rockeries were exposed and hence lacked any greenery and life. What's more the statement of "small hills are made out of rocks." does not mean no use of soil at all in hill construction. Without soil, trees and plants will not be able to subsist. On the other hand, though renovated extensively in Qian Long period of the Qing dynasty, the Mountain Villa with Embracing Beauty is considered as one of the best gardens in the region south of Yangtze for its superb rockery arrangement. In a small piece of land, probably within a stone's throw, the mini-landscape of creek running quietly among secluded deep gorges seems to be a natural wonder. In the Song dynasty, a painting technique called "chopping with axe"（"斧劈法"）was developed in painting rocks and hills. This painting style was adopted in the rocky design of the Mountain Villa with Embracing Beauty, which strikes one with its boldness and force（Page 217）. With winding waterway nurturing the hillsides and rocks, the cliffs come alive with joyful green. Such an artistic perfection in design cannot be achieved if one has neither life experience nor an ability to put his knowledge into play with facility and dexterity. Here one can see the importance of conscientious study and self-cultivation.

Arrangement of a hill in garden design should be based on the topography of an existing terrain. The natural topography is variable, which makes it impossible to follow a fixed format or pattern in arranging hills in a garden. However, any formation and development in nature has its own law to abide by. Therefore, the saying "learning from Creator is more worthwhile than learning from ancients." has a grain of truth. Only by thoroughly understanding this philosophy, can one consummate a good design in garden rockery. In brief, one should initiate the design with a careful analysis of existing landscape. In regards to the previous work done by the ancients, one needs to understand its essence of beauty and try to apply it to today's work. At same time, one should also try to avoid replicating the awkward design done in the past. Unfortunately, this philosophy was not very well followed in Suzhou gardens constructed lately. The majority of these designs were just a jumble of buildings and ponds pieced together without any careful study. A large number of designers fell into the mindset of going by a fixed formula without any iota of innovation. This is especially true of the selection of rocks for stacking. Instead of taking the whole landscape into consideration these people just treat each piece of rock as an antique and rigged up these rocks together without a careful planning of their contrast with plants and their harmony with buildings. As mentioned earlier, the Garden of Pleasance was intended to benefit from the best designs of all the gardens in Suzhou. But the end product is a mere patchwork of structures, which does not signify a success. This is a lesson to today's garden architects.

As to the water in a garden, what is of necessity is to locate a source. A pond without a source is destined to become stagnant. For example, pre-existing lakes and swamps are used to create water surface in the Humble Administrator's Garden. A source of water is found in the Mountain Villa with Embracing Beauty by digging up an underground spring. The spring water is limited in volume but fresh and limpid enough to nurture the plants.

Proper design methods should be applied to create the impression of expansiveness of water surface in spite of a limited garden space. In general, there are two common practices to achieve this effect. The first method is to take advantage of an irregular water surface and sprinkle it with some islets. The islets are all connected via small bridges. With this arrangement, the space over the water is not plainly visible at a glance. The second method is to design with meticulous care of shoreline and upstream sources. The design involves placing various small bays or fjords along the shoreline, which brings about an impression that the water flows from a variety of sources with neither beginning nor ending. It adds a sensation of seclusion by hiding the sources in deep gorges. This effect of illusion is further enhanced by planting reed and wild rice at the upstream sources. However, the latter method is viable for larger gardens.

The building Huo Po Po Di (The Place of Liveliness, 活泼泼地) in the Lingering Garden is a waterside pavilion located at the end of a creek. But a part of water still flows underneath the building. With such an arrangement, the pavilion seems to stand over a never ending creek. The water rockery in Hui Yin Yuan at Nan Xian Zhi Xiang (南显子巷惠荫园) is made from layers of carefully stacked rocks. Water is siphoned into the tunnels formed within the rockery. While pacing upon the stepping stones propped up in the water, one feels like walking along deep gorges. And a hut is placed on the top of the cave. This unique structure found in central Wu region is based on the design of Zhao Yiwang's garden in the South Song era (南宋杭州赵翼王园). The highlight of the garden Cang Lang Pavilion (沧浪亭) is its rockery. There is a waterside veranda on its east side. This veranda rises in elevation so drastically as to offer a bird's eye view of the pond deep below. It is unfortunate the design does not fully achieve the dramatic effect because there is no flowing water in the pond. This type of detail should not be neglected in design in order to achieve consummate effect.

As will be discussed in detail later, bridges in Suzhou gardens are typically close to the water surface. The bridges in the Garden of Cultivation are almost level with the water (Page 215), which sets off the height and steepness of the rocks on the shore. Although the bridges in the Garden of Pleasance are a little bit away from the water, they are sufficiently lower than the surrounding rockeries, which is still acceptable. What would the contrast be when building a bridge over a small creek? It is not hard to imagine why our ancients used stepping stones instead of a bridge. This method seems more in keeping with natural order. There are few examples of waterfall in Suzhou gardens. The waterfall over the building eave found in the Mountain Villa with Embracing Beauty is an exception.

Building structures such as halls, towers, pavilions, kiosks and terraces compose other major elements beside water ponds, creeks, rockeries and hills in Chinese gardens. Gardens in the area south of Yangtze are famed for their tranquility and elegance. The style of building structures in these gardens should be exquisite and in harmony with the surroundings. So their position, terrain and exterior appearance should be taken into consideration. The famous book of Chinese garden construction, *Yuan Ye* (*Craft of Gardens*, 园冶), stated: "At the start of a garden construction, one needs to decide how the hall should be built. The building should match the scene. Facing south is preferred. It is also desirable to plant some trees in the courtyard." This is the principle followed in the construction of the gardens south of Jiangsu, such as the Hall of Distant Fragrance in the Humble Administrator's Garden and the Han Bi Mountain Villa in the Lingering Garden.

As to the positioning of a hall in a garden, there is no strictly defined principles to go by, though some guidelines and advice can be found in several classic books of Chinese garden construction, such as *Yuan Ye*, *Chang Wu Zhi* (*Treatise on Superfluous Things*, 长物志), *Gong Duan Ying Zao Lu* (*Notes of Hall Construction*, 工段营造录). The following are some guidelines quoted from these books. "Form of a building accords with the base on which it is built." "Arrangement of a building should be consistent with its surrounding." "Each aspect of the design should be considered separately to meet its own need." "Each corner is wrought according to its own feature." "A bower should be hidden in a grove. A kiosk should be positioned by water." In summary, one should not fall into a rut in designing a hall. In stead, one should proceed from the construction of a garden in its entirety and use some common sense in hall design.

Take the Humble Administrator's Garden as an example. If we take an aerial view of the garden, we can find that its buildings mix themselves harmoniously with the changing landscape. Their exterior look is elegant and sprightly, no matter what plane views are presented, rectangular, square, polygon, or circle. There is also a rich variety of roof styles, including the gable roof with either flushing purlins or outstanding purlins （硬山，悬山）, or the gable-on-hip roof either on both gable sides or on single gable side （歇山，单面歇山）, or the pyramid-like pointed shape roof （攒尖）. The only exception is the palace-like hip roof, that is to say, a hip roof with a heavy eave structure and a heavy outstanding ridge （庑殿式）. Most roofs in the garden are without the main ridge, making a smooth curving transition at its top line （卷棚式）. The upturned corner eave adopts a "water spraying" style （"水戗发戗"）, which starts the upturning at a low angle and thus looks light. This light touch of corner eave is in line with the intricate decorative border beneath the eaves, as well as the balustrade called "Backrest for King Wu" （吴王靠） slightly curved between the building pillars.

Regarding the elevated structure of buildings, the center section of the Lingering Garden is a good example. Looking eastward from the Osmanthus Fragrance Pavilion, the Winding Stream Tower comes in view with its gable-on-hip roof. Its ground floor is designed like a foundation of a platform, with architrave in parallel to the water and separated from the upper floor. With such an arrangement, this two stages tower does not look overwhelming in height. The locations of the three buildings, namely, the Winding Stream Tower, the Western Tower, and the Refreshing Breeze Pavilion by the Lake, hold off from each other. The roof structures of these three buildings are of the same type but each is slightly distinctive in its own way. Such arrangement matches well with the rockeries and water around them. A contrast of virtual and real visions of architrave and wall is created through the arrangement of brick framed tracery windows casting sparse and light flowery patterns. This is a unique feature found in Suzhou gardens. In particular, with the reflection of all these structures in the water the beauty of the scenery beggars any description. Set off by such a milieu, an aged tree leaning lonely over water seems more sturdy and vigorous.

Another prototype of flexible hall design is witnessed in the Han Bi Mountain Villa and the Pellucid Tower in the Lingering Garden. These two buildings are attached to each other. The roof structure of the Villa is a gable roof with flushing purlins on both sides. The roof structure of the Tower is a hybrid one, a simple gable with flushing perlins on the side attached to the Villa and a gable-on-hip structure on the open side. Usually, this special hybrid roof design would have looked odd when applied to a single independent building. Nevertheless, when applied in this particular case, the two adjacent buildings look more lively instead (Page 159). A similar example is Chang Yuan (The Garden of Free Will, 畅园). This garden is narrow in width. With a pond in the middle, there is little space to build a pavilion. Consequently, a single side gable-on-hip structure is used in the place of waterside pavilion.

On the other hand, there are some cases in which the building design needs to be re-considered. One example is the pavilion at the corner of the east section of the Lingering Garden. A hexagonal pyramid roof structure is used in this case. In order to seek a visually delicate and graceful effect, a "spraying water" style is adopted on its upturning corner eaves. However, the pillars are too thin and too tall to support such a roof structure. Consequently, this pavilion looks out of proportion. The Free Roaring Pavilion and the Delightful Pavilion are both located in the west section of the Lingering Garden. These two pavilions both suffer from disproportions. The Free Roaring Pavilion is small but not delicate. The roof of the Delightful Pavilion seems too heavy though full of variation. Indeed, this is a common problem in recent pavilion construction in Suzhou region. The modern time artisans of Xian Shan generation preferred pavilion structure with overly high roof and thin columns. The exterior look of these pavilions is not necessarily attractive. This modern design is divergent from the Ming style design found in Yi Pu (The Garden of Cultivation, 艺圃), which has a slightly lower roof structure to enhance the stability of a building.

In the last analysis any single building design should be conceived in light of the whole garden arrangement. As to plane variability the design calls for a fluent and smooth flow in a winding pathway as well as a feeling of spaciousness in a densely furnished hall. The size of a building is completely dependant on its need and milieu. It can be as small as one bay or two bays, some times even two and half bays. The Malus Spectabilis Garden Court in the Humble Administrator's Garden is only a two bays court, with the major bay obviously wider than the other. The Worshipping Stone Pavilion in the Lingering Garden is two and half bays wide, among which the half bay is the most intriguing. Only up to this point one can really appreciate the design principles presented in the classic book *Yuan Ye* (*Craft of Gardens*). High and low in profiling, left and right in balancing, virtual and real in contrasting, every aspect is not to be ignored in designing a garden building.

The furnishing and decoration of the following places are among the best in Suzhou gardens. They are Wang Shi Yuan (The Master-of-nets Garden, 网师园), the Flower Hall in Gu's House at Tie Ping Xiang (铁瓶巷顾宅花厅), Cheng's Garden in Xi Bai Hua Xiang (西百花巷程氏园), the Flower Hall in Wu's House in Da Shi Tou Xiang (大石头巷吴宅花厅), the east and west Flower Hall in Ren's House in Tie Ping Xiang (铁瓶巷任宅东西花厅), as well as Wan's Garden in Wang Xi Ma Xiang (王洗马巷万氏园). For additional details on this subject please refer to another book of mine, Collection of Decorative Devices. In particular, Wan's Garden in Wang Xi Ma Xiang is a small garden but its study is well isolated from the rest of the garden, creating a uniquely secluded environment. In regards to some other gardens, there is much room for improvement in their furnishing and decoration. The Lingering Garden is overly complex. It looks coarse and crude with few exquisite designs in between. In the Humble Administrator's Garden, although some fine decorations can be found, most are plain and invariant. This is the result of piecing together only what are available during the restoration and repairs. As far as Yi Yuan (The Garden of Pleasance, 怡园) is concerned, almost all original furnishings have been lost, except the Boat-Like Pavilion, which is among the best in the central Wu region (Page 229).

Verandas are the veins and arteries of a garden. They deserve special attention due to their importance. Double-corridor verandas are quite common in today's Suzhou gardens. This is a kind of veranda with a wall in the middle separating the verandah into two sides. Tracery windows (see my writing *On Tracery Windows* for additional details) are built in this center wall. The two sides of the verandah are available for passage and often referred to as inner or outer corridor respectively. The double-corridor veranda is used mostly as the boundary of an open style garden, such as the one along the water in Cang Lang Pavilion（沧浪亭, Page 261）. It is used also inside a garden as a division in order to increase an illusion of spaciousness or as a transition between the entry and the garden interior. One such example is the double-corridor veranda found in Yi Yuan (The Garden of Pleasance, 怡园, Page 224). Other than the function mentioned above, this double corridor also provides a shade to Sui Han Cao Lu (the Evergreen Thatched Cottage, 岁寒草庐) and Bai Shi Xuan (the Rock Fetish Pavilion 拜石轩). It protects the two buildings from the afternoon sunshine and the strong west wind. On the other hand, the sunlight sieving through the tracery windows of the partition wall makes a fantastic view of intricate lattice patterns.

In general, verandas can be classified according to their location into those on land and those by water. It is also possible to categorize verandas in terms of their shape into winding verandas and straight verandas. The shape of a veranda should avoid being flat and straight. On the other hand the verandas in present day Suzhou gardens are overly windy by design. This type of designs lacks a natural touch and looks stilted. An example of such design is the passage along the north wall of the central section in the Lingering Garden. This veranda is accosted by a brick balustrade, which seems stifling and does not match the decorative border at the top. The placement of columns between stone tablets further adds oppressiveness to the design. A better design is achieved by using hollow bricks for the waterside verandas found in both Ren's House in Tie Ping Xiang（铁瓶巷任宅）and the west side of the Humble Administrator's Garden. For the veranda "A Winding Way in Willow Shades"（柳阴路曲）in the Humble Administrator's Garden, wood balustrade is used on the waterfront side, while Wu Wang Kao (Backrest for King Wu, 吴王靠, a bench style balustrade with straight upward backrest) is placed on the opposite side. This type of veranda design stands to reason (Page 128).

The best waterside veranda can be found in the west side of the Humble Administrator's Garden (Page 132). There is not only an extremely well designed zigzag passage, but also a slight slope. This is just like what is called "floating veranda takes you across the water." in *Yuan Ye* (*Craft of Gardens*, 园冶). The planting of plantain or the setup of a lake stone rockery along selected winding areas of the veranda with small waterside pavilions standing by are some of the highlights in this design (Page 133-134).

Ascending verandas can be found along border walls of various Suzhou gardens. This type of veranda is used sparingly in gardens such as Shi Zi Lin (The Lion Forest Garden, 狮子林), the Lingering Garden, and the Humble Administrator's Garden. It is important to pay attention to the relationship of the veranda's slope with the hill's height. Otherwise, an improperly designed ascending veranda would be top heavy and out of balance. If the space available is limited or uneven in width, it is advisable to use either a single side slope, or combination of single side and two sides slope as needed. This type of design can be found in the west part of the Lingering Garden. The planting of bamboo or plantain, and placement of lake stones along the inside corners of a winding veranda will always result in a spot of wonderful scenery.

Li Dou (李斗) wrote "A wooden structure along with brick walls makes a passage. When the passage is curved along a land slope it becomes a touring passage. If the passage is both curved and bended along its way, it is turned into a winding verandah. When curvature is removed, it makes for an addressed pathway. If two ways run opposite to each other, it is called a paired way. Those for the purpose of passage are walkways. If for wandering, it is named stepping way. When leading to a bamboo grove, it is designated as a bamboo path. When built by the water it is addressed as a waterside verandah. At some places it sits against cliff, with high rail. At other places it stands over the water, with boats flowing beneath. Here and there, it looms in the flower grove. Now and then, it emerges at the shore. At most places it winds among pavilions, towers, and halls, providing an easy passage. A verandah cannot miss its rail. A verandah with a nice rail is just like a beauty wearing a pretty short sleeve which gives prominence to her slender waist. Putting boards over the rail makes a flying-in bench, or what is called Backrest for Beauty (美人靠). The portion of a veranda with wide interior space is called Xuan (轩, a verandah type hall)."

Li Do's writing provides a detailed overview of verandas and serves as a good reference. Nowadays some applications of garden verandah have been found. For example, the veranda lined on both sides with plantain is called a Ba Lang (Plantain-thronged Veranda), while a verandah ringed with willow trees is called a Liu Lang (Willow-sheltered Veranda). These verandas offer verdant shade to keep the passage cool during summer while in winter are bathed in warm sunshine. During ancient times, some verandas were used as a gallery to exhibit paintings. In recent years, some of these verandas are decorated with inlaid stone tablets inscribed with poems on the surface. All these are superb adaptations of ancient verandas.

Bridge is an integral part of a garden. In a garden, a bridge to the water is as significant as a verandah to the road. Clapper bridge is the most common bridge found in Suzhou gardens. This type of bridges can be categorized into several styles including straight bridge, zigzag bridge with three, five or nine bends, and arc shaped bridge. The height of the bridge can be either level with the shore, as found in more recent designs, or lower than the shore and slightly elevated from the water, as found in the older bridges. The bridges following older design are still available in the Garden of Cultivation, Ji Chang Yuan of Wuxi (Garden for Lodging One's Expansive Feelings, 无锡寄畅园), and various gardens in Cangshu (Page 270). The bridges found in the Garden of Pleasance as well as in the ruins of Yan family garden in Mudu also bear resemblance to this design, but they are slightly higher in elevation from the water surface. In my view, the older design is motivated to achieve the following effects. Above all, walking through a bridge at water level gives people a feeling of treading the water. It enhances the perception of vastness of the water and of precariousness of the bridge. Secondly, a lower bridge surface will form a sharp contrast with the height of the surrounding buildings and hills. The steepness of cliffs and buildings seem more impressive. The bridge in the Garden for Lodging One's Expansive Feelings is arranged in a secluded canyon beneath a plateau. Behind the plateau is the transferred scene of Hui Mountain in distance. This arrangement demonstrates the complexity of garden design techniques employed during the Ming dynasty.

On the other hand, clapper bridges are often not deployed very carefully in today's gardens. They are designed without consideration of the surrounding landscape and the size of the bridge. In particular, the balustrade design is incompatible with both the bridge and surrounding environment in terms of its height and style. In some designs, there are even wrought iron balustrades typically found in semi-feudal and semi-colonial times. The bridge in the west section of the Humble Administrator's Garden is such an example.

Regarding some good bridge designs, the small bridges in the Garden of Cultivation and the Leaning against Rainbow Bridge in the Humble Administrator's Garden (Page 96) are among the most admirable. The height of balustrade for the two zigzag bridges in the center section of the Humble Administrator's Garden is also appropriate. However, both bridges are slightly too elevated from the water surface. There is a stone bridge connecting the two islets where the Orange Pavilion and the Snow-Like Fragrant Prunus Mume Pavilion are located respectively. This bridge is made of only one single stone slab without any balustrades (Page 108). This is a primitive but praise-worthy bridge design.

The other style of stone bridges in Suzhou gardens is the small stone arch bridge. They can be found in both the Lion Forest Garden and the Master-of-Nets Garden, but the former is not as good as the latter in design. The reason is obvious. In general the water surface in Suzhou gardens is not very large. Arc bridge usually has a giant and solid body. Placing such a bridge across a small pond makes the view rather oppressive. In the Master-of-Nets Garden, this problem has been solved successfully. The bridge body there is quite small. It is built at the end of the water (Page 219). When viewed from the west, an exquisite reflection of the bridge is clearly visible in the expansive water. And conversely, when looked from the east of the bridge, a moon gate is mirrored in the water. The effect of these two arrangements is similar. In Ban Yuan at Zhong You Ji Xiang (中由吉巷半园), the arc under the bridge is deformed to adapt to the available limited space. As regarding the way across a creek, the book *Yuan Ye* (*Craft of Gardens*, 园冶) said "laying small stepping stones" is a more natural and much better approach than building a bridge. It is unfortunate this style of design is almost lost today.

The road in a garden is noted in the monograph *Provision of Leisure*（清闲供）as "the trail inside a gate must be winding" and "the way beside a house must be extendable to all directions". For example, the trails in the center section of the Humble Administrator's Garden follow the original Ming dynasty architectural arrangement. They are winding in a graceful fashion along the land-shaped profile. There is a clear distinction between primary and auxiliary roads. The trails in Huan Xiu Shan Zhuang（The Mountain Villa with Embracing Beauty, 环秀山庄）are more curvy and winding due to the limited space of the garden. But they are arranged properly and intriguingly. On the other hand, the trails in Shi Zi Lin（The Lion Forest Garden, 狮子林）and Yi Yuan（The Garden of Pleasance, 怡园）are interlaced with each other in a way that is both unnatural and inexplicable. Indeed, any cutesy arrangement in the garden far removed from the nature is not truly artistic and should be avoided.

The pavement is another important aspect in the design of a garden. The pavements of a court front, a trail, or a main road are equally important and need to be considered carefully. In today's Suzhou gardens, bricks are commonly used to pave the main road. Its paving is relatively simple, and easy to form various patterns. Gravel pavements can be used on either road or courtyard ground. It is usually speckled with clay or porcelain discs to form patterns. The pavements made of cobblestone, or cobblestone with porcelain discs are more elaborate and elegant than the pavements using brick or gravel. But they call for more maintenance. The earth filled between cobblestones gets lost after rain wash or road sweeping and hence needs to be refilled regularly. Weeds growing in the earth also need to be removed periodically.

Ice cracking pavement is a special technique featuring the courtyard ground. Two kinds of this pavement are applied in the region south of Jiangsu. The first kind is just to place natural rocks on the ground leaving ice cracking veins around them, such as the pavement in front of the Hall of Distant Fragrance in the Humble Administrator's Garden（Page 152）. It looks natural, but has a difficulty to make the pavement evenly spaced. Another kind is to cut the rocks into slabs and then to match up these irregular slabs while paving, leaving crevices between them, such as the pavements in front of Han Bi Mountain Villa in the Lingering

Garden (Page 198) and Hua Ting (the Flower Hall) in Gu's House at Tie Ping Xiang (Page 241). This is a very difficult paving job. But it gives a solid and neat ground surface. Another related structure is the steps in front of a hall. Piling natural stones to form the steps creates an air of wilderness. A case in point is the kind of steps in front of the Hall of Distant Fragrance in the Humble Administrator's Garden (Page 152) and the Celestial Hall of Five Peaks of the Lingering Garden (Page 173).

Various kinds of walls are erected in Chinese gardens, such as gravel wall, water-milled brick wall, latticed brick wall, and stucco wall. Among them stucco wall is the most common in today's Suzhou gardens. Some times, tiles are put on top of outer wall to form a latticed top portion. In most cases inner walls are inlaid with brick-framed windows or tracery windows. This is commonly referred to as "flowery patterns on white-washed wall". The application of water-milled wall is limited to building structure. Most entrance gate wall is built by laying water-milled bricks in various patterns. A good example is the gate wall of the Humble Administrator's Garden (Page 83). A wall can be divided to upper and lower parts. The lower part is called wall shirt. A short portion just above the wall skirt is called skirt shoulder. In Chinese gardens, gravel wall is mostly found at this portion of the wall. In an ancient house in Shanghai, I saw a very refined wall with ice cracking pattern dotted with plum flower, which is believed to be from the Ming dynasty. In the West Garden (戒幢寺西园), a latticed window wall is built across water, a unique device indeed (Page 252).

Antithetical couplet and horizontal tablet are important and necessary decorative elements in Chinese gardens. Their role is just like what the beard and the eyebrows play for a man. Suzhou is a region where assemble many talented people. Many writers and painters living in this region are involved in garden design. They are looking forward to a place full of idyllic charm comparable to that of natural landscape and famous scenery spots. Or they yearn for a site which can materialize the splendid artistic conception expressed in their poems and paintings. Therefore, they have penned the most elegant and beautiful phrases to describe the objects in a garden, including rockery, creeks, trees, hills, pavilions or halls.

These writings of couplets or tablets epitomize the fair sceneries, triggering people's passion and longing for things of beauty. For example, both the Hall of Distant Fragrance and the Stay and Listen Pavilion in the Humble Administrator's Garden are places for enjoying lotus flowers. The name "Distant Fragrance" originated from the phrase "Faint fragrance wafted from afar（香远益清）", while the name of "Stay and Listen" derived from "Listen to the rain spattering on the remnant leaves of lotus（留得残荷听雨声）". The Osmanthus Fragrance Pavilion in the Lingering Garden and the Malus Spectabilis Garden Court in the Humble Administrator's Garden are both named after the species of trees planted in the surrounding areas. It is expected that a large variety of new literary writings will be produced by those people who get inspired through journeying these sceneries. This is a unique feature of Chinese garden culture. I wish this culture would be preserved as much as possible for our future.

There is a rich collection of antithetical couplets and horizontal tablets in many of the Suzhou gardens. In Yi Yuan（The Garden of Pleasure, 怡园）, there are some antithetical couplets adopted from Song lyrics which express the sceneries extremely well. It is unfortunate that most of them have not survived. As regarding the material used to make couplets and tablets, due to the rough wind the majority of autographs are incised on the board of gingko and then filled with Shi Lv（石绿, a special material used in autograph incision）. Sometimes, the incision is covered by other cold color paint. Incisions on bricks are also seen in some gardens. They look nice and clean. Scripts such as seal script, clerical script, or semi-cursive script are the calligraphy styles used most frequently in couplets and horizontal tablets. This is because of their primitive and simple forms. On the other hand, regular script is rarely used. The Chinese calligraphy carving is an art by itself, and is of the same origin as the Chinese painting. It definitely enhances the beauty of gardens.

The importance of plants in a garden is well understood. In particular, the gardens in the region south of Yangtze are small, and most are enclosed by high walls on all sides. In selecting and cultivating flowers and plants, meticulous care should be taken of their location, sunny or shady, their position, high or low, their rate of growth, quick or slow, their resistance to cold, strong or weak, and their shape, simple or sophisticated. First priority should be given to the arrangement of plants and rockery and their compatibility with the scenery. In Chinese gardens, rockeries and ponds are a miniature of nature. Plants should be arranged with care and be prevented from growing wildly. It is important to limit their size and trim their form routinely. Otherwise, gnarled branches by the water and twisted roots around rocks will make a scene look wizened and spiritless. Once Li Gefei (李格非) said aptly: "Better effort in design can avoid a wizened and worn scene."

In Suzhou, one can find a variety of plants in different gardens. For example, elm is grown mainly as big trees in the Humble Administrator's Garden. In Lingering Garden, gingko is frequently found in the center section while maple is spread all over the hill in its west part. Yi Yuan (The Garden of Pleasance, 怡园) is a small garden. Accordingly, the trees planted there include cassia, pine, and lacebark pine. In particular, the lacebark pine is small in size and assumes a simple shape. This is one of the most precious species of trees for a small garden. Other trees such as pine, plum, palm, or boxwood are also cultivated in Suzhou gardens. These trees tend to have long growth cycles. For a small garden ringed with high walls, its land lies mostly in shadow. This requires planting of cold resistant plants along the wall, such as privet, palm or bamboo. The trees chosen for high hillocks should be of the alpine type such as pine and fir. In the crevices between entrance stone steps shade-tolerant grass is the best choice. With such an arrangement, a garden will flourish with more evergreens than with deciduous trees, providing a constant source of green throughout the seasons and avoiding sterility during fall and winter. Trees such as elm, pagoda tree, and maple should be trimmed annually in order to maintain its delightful and primitive shape. Their root structures are especially adorable. A full-grown tree like elm often presents a picturesque scene by itself, with flourishing verdant twigs grown on twisted branches and a decrepit trunk.

In today's Suzhou gardens, there are in general two ways to plant trees on the top or the slope of a hill. One way is to plant large trees with their ground parts open. This can be found in parts of both the Lingering Garden and the

Humble Administrator's Garden. Thus being done, the beauty of the tree root with rocks around it may be revealed for appreciation. The other way is to plant trees amongst bamboos or shrubs; and the rocks are covered with vines. This is the case with the Garden Cang Lang Ting (The Cang Lang Pavilion, 沧浪亭) as well as part of the Humble Administrator's Garden. It offers a view of lushness and profundity. From my observation, these two ways of tree planting follow different ideas and originate from different painting scripts. The former complies with the style of the Yuan era painter Ni Zan (Yun Lin)(倪瓒［云林］) which is noted for being fresh and elegant, while the latter abides by the style of the Ming era painter Shen Zhou (Shi Tian)(沈周［石田］) which is known for being lush and dark-toned. Regarding the plants in the low land along the shore, it is advisable to choose willow, bamboo, or reed. Shaded area by the wall should be covered with Japanese creeper, sacred bamboo, begonia, and slender bamboo. These types of plants with green leaves and cold colored flowers present an elegant sight. Again they need to be trimmed annually.

The principle of growing flowers and planting trees in a garden is one and the same. Osmanthus and camellia should be planted in shaded areas which receive little direct sun light. They are evergreen plants, with osmanthus flowering in autumn while camellia blooming in early spring. Both types of plants blossom before others. It is enchantingly beautiful when these full-blooming trees are rollicking among arrays of quaint-shaped rockeries. Vines such as wisteria will form a green shade over the pergola when spring begins. The multitudinous flowers sprinkled across a sprawling green look like the luster of jewelry on emerald velvet. Peony is preferably grown on a platform skirted by stone balustrade. This is because growing peony needs abundance of sun shine and high-rise land. The high location naturally enhances the magnificence of peony flower. Traditionally, one prefers to plant magnolia, malus, peony, and osmanthus in the same courtyard since their names rhyme with the words for "jade, dignity, wealth, honor" in Chinese. Though this implication might be inadequate for today's life, the way our predecessors to reconcile the blooming periods and colors of various flowers is still worthy of appreciation. The flowering trees like peach and plum should form a grove to be viewed from afar.

However, these types of trees should only be cultivated in large gardens, such as the Lingering Garden and the Humble Administrator's Garden in Suzhou.

Two principles are followed in arranging trees in Suzhou gardens. The one is to plant the same type of trees to form a grove, such as the pines at the Place of Listening to Waves in Yi Yuan (The Garden of Pleasance, 怡园听涛处), the maples in the west section of the Lingering Garden, as well as the osmanthus in front of the Osmanthus Fragrance Pavilion (闻木樨香轩) in the Lingering Garden. While applying this principle, one should consider the height and the density of trees in relationship to the surrounding environment. The other principle is to cultivate a variety of trees. Whether it is viable or not depends on the orientation of trees and the nature of land. While selecting tree species, one should consider the contrast and balance of colors, the ratio between evergreen and deciduous trees, their blossom seasons, the shape of their leaves, and their postures. The selection of trees is indeed similar to composing a painting. The relationship between trees and rocks needs to be considered too. It is desirable to create a multitude of views along a hill, making every inch of rock scenery lively and alluring. It is also important to see to the consistency between plants and buildings in terms of the latter's forms and colors to set off other's beauty.

Lotus needs to be planted in a pond, but in a limited way. Otherwise, the water surface would be greatly reduced with numerous lotus plants, which hinders the beautiful reflection of buildings. To adorn the water with a sprinkle of lotuses has a greater visual impact. The slim and graceful plants swaying in the middle of the pond is a perfect delight to the eye. In Gu family garden at Kunshan, lotus was reportedly planted by placing the roots underneath stone slabs lining the pond. There are a few holes on the stone slabs allowing the plant to come through. This technique limits the propagation of the plant. Master Liu Dunzhen (刘敦桢) mentioned "In Dong Yuan of Xu family in Nanjing of the Ming era, the lotus was planted by placing the roots in a vat sunk in the pond." That technique achieves the same effect of limiting lotus growth.

The main color theme of the gardens in the region south of Yangtze is characterized by their tranquility and elegance. This is in sharp contrast to the color theme of the royal gardens in northern China, which are famed for their magnificence and splendor. In my analysis, this difference comes from several origins. First of all, the color theme of the gardens in the region south of the Yangtze is related to the cold color theme of the local residence buildings. In this region, the column capital of a building is painted often in either chestnut or non-glossy black, and the decorative border is in dark green. The wall is white in most cases. The sharp contrast between stark black and white is toned down by tracery windows or brick-framed windows which serve as traditional structures. Many spectacular scenes have been created out of this design.

Indeed almost all the Suzhou gardens were originally attached to residential houses for the purpose of providing a quiet place for study and self cultivation. This motivation sets the tone for the cold color scheme. Consequently the design philosophy of Suzhou gardens is obviously different from that of northern royal gardens which was intended to demonstrate the royal power and opulence. In contrast, the scholar-officials in Suzhou showed off their affluence through refined furnishing, meticulous rockery selection as well as exquisite exhibition in their gardens while maintaining a tasteful style of light and cool colors.

The preference to cold color theme in Suzhou gardens resulted also from the influence of paintings in this region. The natural landscape paintings in southern China are composed of strokes of dark shades accompanied by touches of few light colors. This natural and elegant style is imprinted on the mind of scholars, writers, and artists of the region. Its influence leads the garden design away from use of exuberant and glossy gold colors. In addition, since it is extremely hot during summer in the south of the Yangtze River, the rich colors such as red do not fit well. Furthermore, the commoners in the feudal times were not allowed to enjoy the same life style as royal households. The color scheme is consequently limited to those that are light and quiet. One must use a quiet and elegant design to outshine the lavish excess, and use few to outmatch many. This cold color scheme is in harmony with the exterior look of Suzhou gardens: cloudy sky, flourishing trees, exquisite rockeries, as well as the gentle flow of water, which together creates a sense of tranquility and elegance. This is one of the unique features of the gardens in the south of Yangtze region.

Another peculiar feature of all Chinese gardens is that they should be designed in such a way that under any natural conditions, rainy or shiny and at dawn or at dusk, their scenery invariably gives maximum sense of comfort and pleasure. However, without well-planned halls and pavilions, well-made verandas and balustrades, well-arranged trees and flowers in addition to well-wrought rockery and brooks, sceneries can hardly be appreciated in various natural conditions.

Different circumstances can evoke different feelings in the mind of poets or writers. Plantain shaded veranda in summer, sturdy plum and moon light on snow in winter, sunlight and flowers in spring, red knotweed and reed-ringed ponds in fall, these scenes vary with seasons, but are all enjoyable. The rustling of wind through pines, the spattering of rain on wild rice and sweet flags, the swaying flower shadows under the moon, and the fog-cloaked towers, these are all scenes that can elicit different emotions from different people. Garden designers should rely on their profound knowledge in art and literature and base themselves on actual environment to create a garden that can trigger all these human reactions to its sceneries through glamorizing and exaggerating the effect of the most outstanding features. For this purpose, the materialization of a dreamland, they should exert themselves to do whatever they can, laying rocks, raising buildings, channeling water and cultivating plants.

An artist or a man of letters does not indulge in the splendor of spring days only, but has also learnt to enjoy other seasons. They should resort to different ways to ferret out the beauty of each possible circumstance. Watching the swaying flower shadows, they are thinking of what they could do on a whitewashed wall. Listening to the sound of wind, they are imagining the rustling pine trees. Pattering rain revokes pearls jumping and rolling on lotus leaves. Lucid moon reminds them of the moonlight-touched tops of swinging willows. Sunset prompts them to twilight radiating through a tracery window. Year-end cold endears them to the sturdy plum and slender bamboo. They instill their passion in every rock and every plant in order to realize their dreamland. Therefore, in a Chinese garden, every corner is imbued with a sentiment and every side exudes a meaning. It leaves a lasting and pleasant aftertaste. This is all it means for a Chinese garden.

V

I have put together my own observations here to provide readers a reference to Suzhou gardens. It is my belief that the value of gardens in Suzhou should not just be an important and illustrious page in the history of Chinese gardens. It should still be appreciated as a place of recreation and cultivation for the people. The material presented here is instrumental in understanding what we can do to foster the splendid culture tradition we have inherited. There may be some inadequacies in my personal observations. Any criticism or suggestion is welcome.

The initial draft completed by Chen Congzhou at the Architectural History Research Group of Tongji University's Department of Architecture in October of 1956.

苏州园林摄影图录

◆

Photographs of Suzhou Gardens

風月一丘壑　今古幾樓臺

In the heaven's breath,
Under the silvery moonlight,
Lie a hill and a dale defying nature's might.
On the human's earth,
Against age-old plight and slight,
Stand still a few bowers and towers basking in brilliant sunlight.

拙政园大门

Front gate of Zhuo Zheng Yuan
(The Humble Administrator's Garden)

Cloud-scraping rocks.

拙政园自腰门内望
View of Zhuo Zheng Yuan
(The Humble Administrator's Garden)
from Yao Men
(Side gate of the Garden)

Beyond the moon gate is another wonderland.

拙政园自别有洞天东望
Center section of Zhuo Zheng Yuan
(The Humble Administrator's Garden),
viewed through Bie You Dong Tian
(A moon gate with a special name of Another Cave World)

人間春正好 風柳搖曳花纏枝

What on earth a wondrous spring can bring:
Blooming flowers sway and willow branches swing.

拙政园自绣绮亭俯视远香堂倚玉轩

Yuan Xiang Tang (The Hall of Distant Fragrance)
and Yi Yu Xuan (The Bamboo Pavilion) in Zhuo Zheng Yuan
(The Humble Administrator's Garden), viewed from Xiu Qi Ting
(The Paeonia Suffruticosa Pavilion)

临水最相宜 东风吹皱漪

Of a dreamland I am forever fond:
By the peony-skirted pavilion and breeze-creased pond.

拙政园绣绮亭远香堂及倚玉轩

Xiu Qi Ting (The Paeonia Suffruticosa Pavilion), Yuan Xiang Tang
(The Hall of Distant Fragrance) and Yi Yu Xuan (The Bamboo Pavilion)
in Zhuo Zheng Yuan (The Humble Administrator's Garden)

Deep and deep and how much deeper is
the courtyard further along.

拙政园枇杷园入口

Entrance of Pi Pa Yuan (Loquat Garden Court)
in Zhuo Zheng Yuan (The Humble Administrator's Garden)

Through the marble moon gate spring sprightly comes,
Flowering raspberry outshines fading crab-apple blossoms.

拙政园自枇杷园望雪香云蔚亭

Xue Xiang Yun Wei Ting (The Snow-Like Fragrant Prunus Mume Pavilion) in Zhuo Zheng Yuan (The Humble Administrator's Garden), viewed through the entrance Moon Gate of Pi Pa Yuan (Loquat Garden Court)

更蹋最高楼 檻然風光 畫棟飛雲簾捲雨

Stepping up to the top of the tower,
A thrilling sight strikes my eye with power.
From the painted rafters surge churning clouds;
Out of the up-rolled curtains well up sheets of shower.

拙政园绣绮亭及云墙

Xiu Qi Ting (The Paeonia Suffruticosa Pavilion)
and Yun Qiang (The Undulating Wall) in Zhuo Zheng Yuan
(The Humble Administrator's Garden)

浮云弄轻阴　绿杨芳径莺声小

Floating clouds and swaying leaves cast fleeting shadows;
The singing of orioles is wafting in the fragrant willows.

拙政园绣绮亭

Xiu Qi Ting (The Paeonia Suffruticosa Pavilion)
in Zhuo Zheng Yuan (The Humble Administrator's Garden)

During the day the sun shines on the shut window,
Toward the eve the screen hangs still and low.

拙政园玲珑馆及绣绮亭

Ling Long Guan (The Hall of Elegance)
and Xiu Qi Ting (The Paeonia Suffruticosa Pavilion)
in Zhuo Zheng Yuan (The Humble Administrator's Garden)

画阑幽径 又海棠开了 香满衣襟

By the painted rails and along the tranquil trails,
The aroma of crab-apple blossoms suffuses your folds and frills.

拙政园海棠春坞入口

Entrance of Hai Tang Chun Wu
(The Malus Spectabilis Garden Court) in
Zhuo Zheng Yuan (The Humble Administrator's Garden)

深院静 小庭空

In the spacious yard all is still and quiet;
In the small pavilion no one is in sight.

拙政园海棠春坞

Hai Tang Chun Wu (The Malus Spectabilis Garden Court)
in Zhuo Zheng Yuan (The Humble Administrator's Garden)

The day is sluggish in passing;
The crab-apple blossoms are found withering.

拙政园海棠春坞

Hai Tang Chun Wu (The Malus Spectabilis Garden Court)
in Zhuo Zheng Yuan (The Humble Administrator's Garden)

A yonder pavilion.

拙政园海棠春坞漏窗

Lattice Window in Hai Tang Chun Wu (The Malus Spectabilis Garden Court) of Zhuo Zheng Yuan (The Humble Administrator's Garden)

水潺潺，花片片，画桥看落絮，浓绿交加

A stream is gurgling on and on;
Colored petals keep parachuting down and down.
Watching on the bridge catkins floating around,
The green turning greener I've ever found.

拙政园倚虹桥及海棠春坞背面

Yi Hong Qiao (Leaning against Rainbow Bridge)
and the back of Hai Tang Chun Wu (The Malus Spectabilis Garden Court)
in Zhuo Zheng Yuan (The Humble Administrator's Garden)

千林未綠 憑闌淺畫成圖

Spring has not yet tinged the woods green;
I lean on the rail to sketch its quiet scene.

拙政园自倚虹亭西望

Center section of Zhuo Zheng Yuan (The Humble Administrator's Garden),
viewed from Yi Hong Ting (Leaning against Rainbow Pavilion)

柳陰深處 斜陽卻在闌干

The oblique twilight lingers on the winding rail;
The recess of willow grove remains outside its pale.

拙政园自倚虹亭望远香堂倚玉轩背部

Yuan Xiang Tang (The Hall of Distant Fragrance) and back of Yi Yu Xuan
(The Bamboo Pavilion) in Zhuo Zheng Yuan (The Humble Administrator's Garden),
viewed from Yi Hong Ting (Leaning against Rainbow Pavilion)

A glimpse of beauty through the ring-shaped moon gate.

拙政园梧竹幽居

Wu Zhu You Ju (The Secluded Pavilion of Firmiana Simplex and Bamboo)
in Zhuo Zheng Yuan (The Humble Administrator's Garden)

玉洞鳴春

The Moon gate echoed with spring.

拙政园梧竹幽居望绿漪亭

Lu Yi Ting (The Green Ripple Pavilion) in Zhuo Zheng Yuan
(The Humble Administrator's Garden), viewed from Wu Zhu You Ju
(The Secluded Pavilion of Firmiana Simplex and Bamboo)

杨柳绿成阴　流水溅·照影

Poplars and willows grow lush and green,
Sprinkling the mini-ripples with verdant sheen.

拙政园绿漪亭望三曲桥

San Qu Qiao (Three Fold Zigzag Bridge) in Zhuo Zheng Yuan
(The Humble Administrator's Garden), viewed from Lu Yi Ting
(The Green Ripple Pavilion)

West of long corridor, long and color-rich,
There lies a red-painted mini-bridge.

拙政园自绿漪亭望梧竹幽居

Wu Zhu You Ju (The Secluded Pavilion of Firmiana Simplex and Bamboo)
in Zhuo Zheng Yuan (The Humble Administrator's Garden), viewed from Lu Yi Ting
(The Green Ripple Pavilion)

亂山深處水濚洄

Whirling water winding along the wild deep dale.

拙政园梧竹幽居

Wu Zhu You Ju (The Secluded Pavilion of Firmiana Simplex and Bamboo)
in Zhuo Zheng Yuan (The Humble Administrator's Garden)

我见青山多妩媚 只今明月费招邀

The green hills are so bewitching;
The silvery moon is so beguiling;
Both vie with each other for me inviting.

拙政园待霜亭及梧竹幽居

Dai Shuang Ting (The Orange Pavilion) and Wu Zhu You Ju
(The Secluded Pavilion of Firmiana Simplex and Bamboo) in Zhuo Zheng Yuan
(The Humble Administrator's Garden)

天寛水寛　風靜林還靜

The wide water mirrors the wide sky.
The still trees stand in the still air.

拙政园自远香堂望雪香云蔚亭

Xue Xiang Yun Wei Ting
(The Snow-Like Fragrant Prunus Mume Pavilion)
in Zhuo Zheng Yuan (The Humble Administrator's Garden),
viewed from Yuan Xiang Tang (The Hall of Distant Fragrance)

翠陰春晝永　斜橋曲水小軒楣

Verdant shades with the long spring day stay;
Swelling streams under the arched bridge flow all the way,
And around the petty curtained bower joyfully play.

拙政园自雪香云蔚亭下西南望倚玉轩香洲

Yi Yu Xuan (The Bamboo Pavilion) and Xiang Zhou
(The Fragrant Isle) in Zhuo Zheng Yuan (The Humble Administrator's Garden),
viewed from Xue Xiang Yun Wei Ting (The Snow-Like Fragrant Prunus Mume Pavilion)

绕墙深丛竹 花气烘人尚暖

Along the wall a hedge of bamboo treads;
From bunches of flowers scented warmth spreads.

拙政园云墙及铺地

Yun Qiang (The Undulating Wall)
and the patterned pavement in Zhuo Zheng Yuan
(The Humble Administrator's Garden)

小径红稀　芳郊绿遍

Along the trails red flowers grow sparse;
Across the fields far and wide spread green grass.

拙政园曲径小桥

Winding trail and pedestrian stone bridge in
Zhuo Zheng Yuan (The Humble Administrator's Garden).

(译者注：这是连接待霜亭和雪香云蔚亭所在的两个小岛的小桥。
Note by translator: This is the bridge connecting the two islets where the
Orange Pavilion and the Snow-Like Fragrant Prunus Mume Pavilion are located respectively.)

看檻曲縈紅　塔夕飛翠　惟有玉闌知

The winding rails are entangled with red rose;
The upturned eaves are hanging emerald in the air.
Of all these who on earth knows?
Only the jade balustrade is aware.

拙政园自玉兰堂北望见山楼

Jian Shan Lou (The Mountain-In-View Tower) in Zhuo Zheng Yuan
(The Humble Administrator's Garden), viewed from Yu Lan Tang (The Magnolia Hall)

落梅亭樹香　芳草池塘綠

The pavilion permeated with the fragrance of fallen plum blossom;
The pond painted emerald by the grass tender and lithesome.

拙政园自别有洞天望见山楼及五曲桥
Jian Shan Lou (The Mountain-In-View Tower)
and Wu Qu Qiao (Five Fold Zigzag Bridge) in Zhuo Zheng Yuan
(The Humble Administrator's Garden), viewed from Bie You Dong Tian
(A moon gate with a special name of Another Cave World)

流水畫橋畔　雨餘芳草斜陽

Under the rainbow bridge gently runs the silvery stream;
Pearly drops of rain on the leaves sparkle in the slanting sun's gleam.

拙政园见山楼侧面

Jian Shan Lou (The Mountain-In-View Tower)
in Zhuo Zheng Yuan (The Humble Administrator's Garden)

長虹臥碧 斜陽低畫柳如煙

Over the grand green land arches the gorgeous rainbow;
In the oblique twilight wisp the misty branches of willow.

拙政园见山楼侧面及五曲桥

Wu Qu Qiao (Five Fold Zigzag Bridge) and a corner of
Jian Shan Lou (The Mountain-In-View Tower) in Zhuo Zheng Yuan
(The Humble Administrator's Garden)

煙外好花紅淺淡　檻前疊石翠參差

Beyond the mist, lovely flowers don the red of light shade;
Outside the rail, thickets of bush dot the rocks artfully laid.

拙政园见山楼背面

Backside of Jian Shan Lou (The Mountain-In-View Tower)
in Zhuo Zheng Yuan (The Humble Administrator's Garden)

煙掃畫樓出 曲々闌干曲々池

As the fog lifts, a multi-color bower looms large and near;
A winding rail runs along a curved pond lucid and clear.

拙政园见山楼背面爬山廊

Roofed walkway at the backside of Jian Shan Lou
(The Mountain-In-View Tower) in Zhuo Zheng Yuan
(The Humble Administrator's Garden)

小院迴廊春已滿　水閣無塵午晝長

Along the winding veranda spring is everywhere in sight;
Upon the spotless water bower shines brilliant mid-day.

拙政园见山楼下

Ground level of Jian Shan Lou (The Mountain-In-View Tower)
in Zhuo Zheng Yuan (The Humble Administrator's Garden)

With zigzag rails the lengthy corridor is sieged;
By greenish yellow the slender willow twigs are tinged.

拙政园柳阴路曲长廊

Verandah of Liu Yin Lu Qu (A winding way with willow shades)
in Zhuo Zheng Yuan (The Humble Administrator's Garden)

高閣對橫塘 水亭未雨涼先覺

The bower stands tall over an expanse of pond;
The rain breathes coolness before it comes around.

拙政园香洲

Xiang Zhou (The Fragrant Isle) in
Zhuo Zheng Yuan (The Humble Administrator's Garden)

暗香来水阁 花枝破萼柳梢青

Around the bower faint fragrance flops;
Sprouting buds green willow's tips and tops.

拙政园香洲

Xiang Zhou (The Fragrant Isle) in
Zhuo Zheng Yuan (The Humble Administrator's Garden)

绿杨楼阁 芳草池塘

The pavilion is fenced by lines of willow lush and luxuriant;
The pond is flanked by carpets of grass perfuming and pliant.

拙政园自柳阴路曲望倚玉轩及香洲

Yi Yu Xuan (The Bamboo Pavilion) and Xiang Zhou
(The Fragrant Isle) in Zhuo Zheng Yuan (The Humble Administrator's Garden),
viewed from Liu Yin Lu Qu (A winding way with willow shades)

秋风清 秋月明 竹外松边 秋景如画

The autumn air is fresh;
The autumn moon is bright.
The pine trees nestle in the bamboo bush;
So infatuating the scene is in sight.

拙政园小飞虹及一庭秋月啸松风亭
Xiao Fei Hong (The Small Flying Rainbow Bridge) and
Yi Ting Qiu Yue Xiao Song Feng Ting (Autumn Moon and Windy Pine Pavilion)
in Zhuo Zheng Yuan (The Humble Administrator's Garden)

濮邊照影行　天在清溪底　天上有行雲　人在行雲裏

The crowd on the land cast their shadow in the pond;
The cloud in the air in silvery water is found.
In the sky floats the cloud;
In the cloud move the crowd.

拙政园自小沧浪北望荷风四面亭及见山楼

He Feng Si Mian Ting (The Pavilion in Lotus Breezes) and
Jian Shan Lou (The Mountain-In-View Tower) in Zhuo Zheng Yuan
(The Humble Administrator's Garden), viewed from Xiao Cang Lang
(The Small Cang Lang)

花暗水房春

Waterside pavilion delights in the shade of spring flowers.

拙政园一庭秋月啸松风亭

Yi Ting Qiu Yue Xiao Song Feng Ting
(Autumn Moon and Windy Pine Pavilion) in
Zhuo Zheng Yuan (The Humble Administrator's Garden)

A solitary man wades through petals fallen;
A pair of swallows hovers on drizzly breeze arisen.

拙政园得真亭

De Zhen Ting (The True Nature Pavilion)
in Zhuo Zheng Yuan (The Humble Administrator's Garden)

有此清凉界　落红铺往水平池

What a cool place!
The path is paved with fallen petals;
The pond brims over with silver ripples.

拙政园倚玉轩

Yi Yu Xuan (The Bamboo Pavilion) in
Zhuo Zheng Yuan (The Humble Administrator's Garden)

A tree-skirted pond afloat with fallen flowers.

拙政园荷风四面亭

He Feng Si Mian Ting (The Pavilion in Lotus Breezes)
in Zhuo Zheng Yuan (The Humble Administrator's Garden)

春透水波明 池上楼台堤上路

Vernal breeze caresses the pond rippling but clear;
Water-sieged bower watches over the dyke near and dear.

拙政园荷风四面亭及见山楼

He Feng Si Mian Ting (The Pavilion in Lotus Breezes) and
Jian Shan Lou (The Mountain-In-View Tower) in Zhuo Zheng Yuan
(The Humble Administrator's Garden)

游絲捲晴晝　斜陽微放柳梢明

Lithe and wispy twigs coil and swing in the sky azure;
Oblique rays of twilight shine upon the willow tops to allure.

拙政园荷风四面亭

He Feng Si Mian Ting (The Pavilion in Lotus Breezes)
in Zhuo Zheng Yuan (The Humble Administrator's Garden)

晴景初升风细细

The day dawns fair and bright,
With breeze blowing soft and light.

拙政园柳阴路曲长廊（未修理前）
Verandah of Liu Yin Lu Qu (A winding way with willow shades),
before repair, in Zhuo Zheng Yuan (The Humble Administrator's Garden)

日影篩閨角

In the sky the sun holds sway;
Behind the fence the shadow shies away.

拙政园玉兰堂前

Front of Yu Lan Tang (The Magnolia Hall)
in Zhuo Zheng Yuan (The Humble Administrator's Garden)

啼莺声在绿阴中 无处觅残红

In the shade verdant, warblers are singing with melodious sound;
In the woods around, remnant petals can no longer be found.

拙政园别有洞天

Bie You Dong Tian (A moon gate with a special name of Another Cave World)
in Zhuo Zheng Yuan (The Humble Administrator's Garden)

A fog-cloaked tower.

拙政园自别有洞天东望五曲桥

Wu Qu Qiao (Five Fold Zigzag Bridge) in Zhuo Zheng Yuan
(The Humble Administrator's Garden), viewed from Bie You Dong Tian
(A moon gate with a special name of Another Cave World)

秋水長廊 水石間

A long-winded corridor meanders its way in fall fair;
A nearby stream is gurgling among the boulders bare.

拙政园水廊

Shui Lang (Winding Verandah over the Water)
in Zhuo Zheng Yuan (The Humble Administrator's Garden)

Graceful bamboo in a harmony with grotesque bowlder.

拙政园水廊自北向南望

Shui Lang (Winding Verandah over the Water) in
Zhuo Zheng Yuan (The Humble Administrator's Garden),
viewed from north end

臺榭繞摩芳　芭蕉籠碧砂

Bowers and towers wreathed with fragrant flowers waving;
Steps and terraces shrouded by pliable plantain roving.

拙政园水廊自南向北望

North end of Shui Lang (Winding Verandah over the Water)
in Zhuo Zheng Yuan (The Humble Administrator's Garden)

萬條楊柳風 高拂樓臺低映水

Myriads of slender twigs of willow
In the vernal wind sway and flow,
Touching tenderly the tower and
Swinging shreds of shade on the pond shallow.

拙政园水廊及倒影楼

Shui Lang (Winding Verandah over the Water) and
Dao Ying Lou (The Tower of Reflection) in Zhuo Zheng Yuan
(The Humble Administrator's Garden)

The moonshine sneaks up the steps of a yard profound,
Shedding stripes of shadow on the railed ground.

庭院無人月上階 滿地闌干影

拙政园拜文揖沈之斋（倒影楼下层）
Bai Wen Yi Shen Zhi Zhai (The Room for Worshipping Wen and Shen,
ground floor of Dao Ying Lou [The Tower of Reflection]) in Zhuo Zheng Yuan
(The Humble Administrator's Garden)

水館風亭

A pavilion on wind-wrinkled water.

拙政园拜文揖沈之斋（倒影楼下层）望宜两亭

View of Yi Liang Ting (The Good for Both Families Pavilion)
through a window in Bai Wen Yi Shen Zhi Zhai (The Room for Worshipping Wen
and Shen, ground floor of Dao Ying Lou [The Tower of Reflection]) in Zhuo Zheng Yuan
(The Humble Administrator's Garden)

日長深院悄 起尋花影步迴廊

In the quiet, sun-bathed yard and along the winding corridor,
I ramble in leisure, looking for flowers in their splendor.

拙政园三十六鸳鸯馆东面入口

East entrance of San Shi Liu Yuan Yang Guan
(The Hall of Thirty-six Pairs of Mandarin Ducks)
in Zhuo Zheng Yuan (The Humble Administrator's Garden)

Full-bloomed roses fiddling with the window,
Diffusing into the hall air fresh and mellow.

拙政园三十六鸳鸯馆东面入口
East entrance of San Shi Liu Yuan Yang Guan
(The Hall of Thirty-six Pairs of Mandarin Ducks)
in Zhuo Zheng Yuan (The Humble Administrator's Garden)

已残芳树　犹缀馀英　深院莺啼人静

The trees are weighed down with age;
But their elegance and fragrance remain.
The yard is tranquil and large;
But the liquid melodies of orioles none can contain.

拙政园三十六鸳鸯馆西面入口

West entrance of San Shi Liu Yuan Yang Guan
(The Hall of Thirty-six Pairs of Mandarin Ducks)
in Zhuo Zheng Yuan (The Humble Administrator's Garden)

水底明霞十頃光 天教鋪錦襯鴛鴦

Twilight sets ripples sparkling for miles and miles around;
A pair of loverbirds are sporting on a wavy carpet of a pond.

拙政园留听阁与三十六鸳鸯馆
Liu Ting Ge (The Stay and Listen Pavilion) and
San Shi Liu Yuan Yang Guan (The Hall of Thirty-six Pairs of Mandarin Ducks)
in Zhuo Zheng Yuan (The Humble Administrator's Garden)

春昼初长

Spring day has come to stay,
And loath to go away.

拙政园三十六鸳鸯馆
San Shi Liu Yuan Yang Guan
(The Hall of Thirty-six Pairs of Mandarin Ducks) in
Zhuo Zheng Yuan (The Humble Administrator's Garden)

日长风静 花影闲相照

The day is spreading its splendor;
The air stirs no longer;
The flowers and their shades fondle each other.

拙政园扇亭山石
Rockeries at Shan Ting (The Fan Pavilion, also called
"With Whom Shall I Sit?" Pavilion) in Zhuo Zheng Yuan
(The Humble Administrator's Garden)

The petty pavilion cuddles up in misty willows;
The soothing stream gently and quietly flows.

拙政园笠亭

Li Ting (The Indus Calamus Pavilion)
in Zhuo Zheng Yuan (The Humble Administrator's Garden)

倚藤臨水 步屧登山

What a delight!
Leaning on the coiling vine by the waterside,
Or ascending the green hill with springy stride.

拙政園笠亭

Li Ting (The Indus Calamus Pavilion)
in Zhuo Zheng Yuan (The Humble Administrator's Garden)

小阁枕清流 风定波平花映水

The bower astride the brook, crystal and purling;
The water without rippling and the wind subsiding;
Upside down in the water mirror flowers blooming.

拙政园扇亭及三十六鸳鸯馆

Shan Ting (The Fan Pavilion, also called
"With Whom Shall I Sit?" Pavilion) and San Shi Liu Yuan Yang Guan
(The Hall of Thirty-six Pairs of Mandarin Ducks) in Zhuo Zheng Yuan
(The Humble Administrator's Garden)

花間亭館柳間門

Blossom-hemmed bower and willow-wreathed gate.

拙政园自扇亭望三十六鸳鸯馆

San Shi Liu Yuan Yang Guan (The Hall of Thirty-six Pairs of Mandarin Ducks) in Zhuo Zheng Yuan (The Humble Administrator's Garden), viewed from Shan Ting (The Fan Pavilion, also called "With Whom Shall I Sit?" Pavilion)

With whom to share such a splendid night?
The crescent moon, cool breeze and me on site.

拙政园扇亭窗

Window in Shan Ting (The Fan Pavilion)
in Zhuo Zheng Yuan (The Humble Administrator's Garden).
The Chinese characters on the inscribed plaque hung above the window bear
the other name of the Pavilion, "With Whom Shall I Sit?" Pavilion.

紅薇影轉晴窗畫

The red roses swinging on a day blue and bright;
Their shadows swaying on the window with delight.

拙政园留听阁

Liu Ting Ge (The Stay and Listen Pavilion)
in Zhuo Zheng Yuan (The Humble Administrator's Garden)

In front of the steps, a riot of flowers grows sun-bright;
Under the bridge, a ribbon of a stream gurgles day and night.

拙政园塔影亭

Ta Ying Ting (The Pagoda Reflection Pavilion)
in Zhuo Zheng Yuan (The Humble Administrator's Garden)

深院宇 小阑干 杨柳外无边春色

Delicate banisters stretch deep into the tranquil garden;
Beyond willows vernal scenery spreads onto the horizon.

拙政园曲廊

A winding verandah in
Zhuo Zheng Yuan (The Humble Administrator's Garden)

乱竹分幽径 芳草到横塘

A trail snakes through bamboo wild and rank;
Lush and sweet grass flourishes on the pond's bank.

拙政园三曲桥

Three Fold Zigzag Bridge in
Zhuo Zheng Yuan (The Humble Administrator's Garden)

花影婆娑清晝永

Flowers dance gracefully on the ground;
With flecks of shadow swirling around.

拙政园远香堂踏跺

Step stones in front of Yuan Xiang Tang
(The Hall of Distant Fragrance) in Zhuo Zheng Yuan
(The Humble Administrator's Garden)

Leaning pines break out by the steps of stone;
In the shade of cliffs withered moss is overgrown.

拙政园远香堂前树根

Tree roots by Yuan Xiang Tang
(The Hall of Distant Fragrance) in Zhuo Zheng Yuan
(The Humble Administrator's Garden)

素壁寫歸來　青山逸不住

I pen down my will of homecoming on the wall;
That not even huge green mountains can stall.

留园古木交柯前廊

The Sedan-Chair Hall in front of Gu Mu Jiao Ke
(The Intertwined Old Trees) in Liu Yuan (The Lingering Garden)

半窗晴日水痕收

The window is half ablaze with sun shining;
The water sparkles but without any rippling.

留园绿荫

Lu Ying (The Green Shade Pavilion)
in Liu Yuan (The Lingering Garden)

朱阑聊掩映 夕阳低户水当楼

Myriad rails set each other off with their radiant vermilion,
The setting sun shines the cottages and waterside pavilion.

留园明瑟楼及绿荫

Ming Se Lou (The Pellucid or The Pure Freshness Tower)
and Lu Ying (The Green Shade Pavilion) in Liu Yuan (The Lingering Garden)

長愛朱闌干影 芙蓉秋水開時

Leaning over the emerald balustrade in fair autumn,
I enjoy the silvery moonshine and the lotus blossom.

留园涵碧山房

Han Bi Shan Fang (The Han Bi [Be imbued with the green]
Mountain Villa) in Liu Yuan (The Lingering Garden)

寶臺臨砌　有情花影閣干

A graceful bower towers over terraced trails,
Caressed by sentient flowers and hugged by love-sick rails.

留园凉台

Liang Tai (Terrace at the north of Han Bi Shan Fang
[The Han Bi [Be imbued with the green] Mountain Villa])
in Liu Yuan (The Lingering Garden)

曲折畫闌　玲瓏樓閣

Ornate rails skirting exquisite towers.

留园涵碧山房及明瑟楼

Han Bi Shan Fang (The Han Bi [Be imbued with the green] Mountain Villa) and Ming Se Lou (The Pellucid or The Pure Freshness Tower) in Liu Yuan (The Lingering Garden)

樓臺競妝點　荼䕷開遍柳花飛

Bowers and towers are decked with charm profound,
Setting tumi flowers blooming all over the ground,
And sending fluffy catkins flying about and around.

留园自凉台东北望远翠阁及汲古得绠处等

Yuan Cui Ge (The Distant Green Tower) and
Ji Gu De Geng Chu (The Study of Enlightenment) in Liu Yuan
(The Lingering Garden), viewed from Liang Tai (Terrace at the north of
Han Bi Shan Fang [The Han Bi (Be imbued with the green) Mountain Villa])

凉风生玉宇 梧桐吹下新秋

Breeze rises cool on nature's call;
Leaves desert trees of parasol.
All harbingers of imminent fall.

留园明瑟楼

Ming Se Lou (The Pellucid or The Pure Freshness Tower)
in Liu Yuan (The Lingering Garden)

The dainty to tower and the painted pavilion,
Pair off to charm the spring-filled garden.

留园自明瑟楼北望可亭

Ke Ting (The Passable Pavilion)
in Liu Yuan (The Lingering Garden), viewed from
Ming Se Lou (The Pellucid or The Pure Freshness Tower)

The beauty of the stream-ringed hillock is beyond compare,
With its rain-cleaned grass, radiant sunset and fresh air.

豁山无限好　雨馀芳草斜阳

留园闻木樨香轩

Wen Mu Xi Xiang Xuan (The Osmanthus Fragrance Pavilion)
in Liu Yuan (The Lingering Garden)

暝烟带树 满庭秋色木樨花

Flimsy mist wafting among the trees at dusk;
Osmanthus rollicking in the autumn air brisk.

留园闻木樨香轩曲廊
Winding verandah at Wen Mu Xi Xiang Xuan
(The Osmanthus Fragrance Pavilion) in Liu Yuan (The Lingering Garden)

Along the long-winding corridor I found myself strolling,
While the yard is quiet and the sun is still lingering.

留园闻木樨香轩曲廊

Winding verandah at Wen Mu Xi Xiang Xuan
(The Osmanthus Fragrance Pavilion) in Liu Yuan (The Lingering Garden)

小院迴廊　暖日明花柳

The corridor of the yard is winding around;
Sun-lit fair flowers everywhere abound.

留园曲廊

A winding verandah in Liu Yuan
(The Lingering Garden)

Dainty flowers nestling by the cute tower.

留园远翠阁

Yuan Cui Ge (The Distant Green Tower)
in Liu Yuan (The Lingering Garden)

小楼春色裏

A tower bathed in spring.

留园自五峰仙馆望远翠阁

Yuan Cui Ge (The Distant Green Tower)
in Liu Yuan (The Lingering Garden), viewed from
Wu Feng Xian Guan (The Celestial Hall of Five Peaks)

A pavilion perching over steps.

留园濠濮亭

Hao Pu Ting (The Hao Pu Pavilion)
in Liu Yuan (The Lingering Garden)

瓊樓玉宇

A marble house with a jade-like hall.

留园曲谿楼西楼及清风池馆

From right to left: Qu Xi Lou
(The Winding Stream Tower), Xi Lou
(Western Tower) and Qing Feng Chi Guan
(The Refreshing Breeze Pavilion by the Lake) in
Liu Yuan (The Lingering Garden)

響屧廊深　午陰嘉樹清圓

Wooden slippers echoing from the depth of the long corridor;
Lush trees frolicking with their shade in the noon's splendor.

留园曲谿楼

Qu Xi Lou (The Winding Stream Tower)
in Liu Yuan (The Lingering Garden)

琐牕淡、花影薄

The pale shadow of sun-bright
blossoms swaying on the window.

留园西楼

Xi Lou (Western Tower)
in Liu Yuan (The Lingering Garden)

Slim bamboo and sturdy pine
Stand by the steps of jade nice and fine.

留园五峰仙馆前

Front of Wu Feng Xian Guan
(The Celestial Hall of Five Peaks)
in Liu Yuan (The Lingering Garden)

畫堂春滿

Spring-sprayed art gallery.

留园五峰仙馆内部
Interior of Wu Feng Xian Guan
(The Celestial Hall of Five Peaks)
in Liu Yuan (The Lingering Garden)

萧疏竹影 乱石穿空

Sparse bamboo shedding slender shadow;
Jagged rocks scraping the sky blue and hollow.

留园五峰仙馆西侧

West side of Wu Feng Xian Guan
(The Celestial Hall of Five Peaks)
in Liu Yuan (The Lingering Garden)

竹色苔香小院深

Jade-green bamboo and sweet-smelling moss
In the depth of the yard grow lush with gloss.

留园五峰仙馆东侧小院

Small court yard at the east side of Wu Feng Xian Guan
(The Celestial Hall of Five Peaks) in Liu Yuan (The Lingering Garden)

Shreds of clouds sail over the small garden,
Shedding fitful drizzle or shafts of light of heaven.

留园汲古得绠处
Ji Gu De Geng Chu
(The Study of Enlightenment)
in Liu Yuan (The Lingering Garden)

小阁无灯月侵窗

Moonlight creeping through the
window of an unlit attic.

留园还我读书处
Huan Wo Du Shu Chu
(The Return-to-Read Study)
in Liu Yuan (The Lingering Garden)

Autumn bamboo serenading by the window.

留园揖峰轩内

Interior of Yi Feng Xuan
(The Worshipping Stone Pavilion)
in Liu Yuan (The Lingering Garden)

閒院深迷　雕欄巧護

The serene yard like a maze is deftly laid,
And skirted with delicacy by carved balustrade.

留园揖峰轩曲廊
Winding verandah at Yi Feng Xuan
(The Worshipping Stone Pavilion) in
Liu Yuan (The Lingering Garden)

半窗晴翠

Half-veiled verdant bamboo foliage.

留园揖峰轩前小景

Scene in front of Yi Feng Xuan
(The Worshipping Stone Pavilion) in
Liu Yuan (The Lingering Garden)

深竹户 小山房 浓绿交阴 芭蕉骤雨

Shrouded in the groves deep,
Nestled in the hill steep,
A homestead is half-veiled in shades of green,
And overgrown with palm pattered by showers of rain.

留园揖峰轩前小院

Court yard in front of Yi Feng Xuan
(The Worshipping Stone Pavilion) in
Liu Yuan (The Lingering Garden)

日麗風和春晝長

The sun shines bright,
The breeze caresses light,
The day of spring lingers longer,
And stays shorter the night.

留园揖峰轩前

Front of Yi Feng Xuan
(The Worshipping Stone Pavilion)
in Liu Yuan (The Lingering Garden)

Flower-scented grand hall.

留园林泉耆硕之馆内部

Interior of Lin Quan Qi Shuo Zhi Guan
(The Old Hermit Scholars' Hall) in Liu Yuan
(The Lingering Garden)

閒庭更與天寬尢　兩峰挐攫尚寒

The courtyard lies tranquil under azure sky;
Into cold air shoot up two giant rocks nearby.

留园冠云峰

Guan Yun Feng
(The Cloud-Capped Peak)
in Liu Yuan (The Lingering Garden)

From lush trees stands out the jade-green house,
Hugged by antique walls coated with shiny brown moss.
Isn't it Eden in the world mundane?
One can't but feel at a loss.

留园林泉耆硕之馆前

Front of Lin Quan Qi Shuo Zhi Guan
(The Old Hermit Scholars' Hall) in Liu Yuan
(The Lingering Garden)

寒却园林春已透

Spring prevails in the park's every corner,
Vanquishing the barren and bleak winter.

留园冠云亭
Guan Yun Ting
(The Cloud-Capped Pavilion)
in Liu Yuan (The Lingering Garden)

翠楼十二阑干曲

Rail-rimmed tortuous verandah prostrate
afore the emerald tower.

留园冠云楼前曲廊

Winding verandah in front of Guan Yun Lou
(The Cloud-Capped Tower) in Liu Yuan (The Lingering Garden)

客行花径曲 幾多綠意紅情

Pacing along the petal-paved sprawling pathway;
Touched by the gaudy colors on sentient display.

留园冠云楼前曲廊

Winding verandah in front of Guan Yun Lou
(The Cloud-Capped Tower) in Liu Yuan (The Lingering Garden)

亭亭

Graceful posture.

留园东园一角亭
A pavilion at the corner of east
section of Liu Yuan (The Lingering Garden)

玉立

Delicate beauty.

留园岫云峰
Xiu Yun Feng
(The Mountainous Cloud Peak)
in Liu Yuan (The Lingering Garden)

A thatch of yellow day lily,
A bunch of dainty bamboo,
A patch of plantain.

留园瑞云峰

Rui Yun Feng (The Auspicious Cloud Peak)
in Liu Yuan (The Lingering Garden)

玉宇净无尘

Purified universe.

留园佳晴喜雨快雪之亭槅扇

Partition Frames in Jia Qing Xi Yu Kuai Xue Zhi Ting
(The Good-For-Farming Pavilion) in Liu Yuan (The Lingering Garden)

倦鳥投林雲返岫

Weary of flying, birds flit back to the nests;
Languid of hovering, clouds hasten home around the crests.

留园舒啸亭

Shu Xiao Ting (The Free Roaring Pavilion)
in Liu Yuan (The Lingering Garden)

古木蒼煙無限意

Misty wild woodland revoking
no end of imagination.

留园至乐亭

Zhi Le Ting (The Delightful Pavilion)
in Liu Yuan (The Lingering Garden)

風梳萬縷亭前柳

Breeze-brushed branches of willow in spring,
Before the bower frivolously sway and swing.

留园西部曲廊

A winding verandah in western
portion of Liu Yuan (The Lingering Garden)

春意随人

Spring follows where the heart goes.

留园西部云墙
Yun Qiang (The Undulating Wall)
in western portion of Liu Yuan (The Lingering Garden)

春水綠波花影外　粉牆丹桂柳煙中

Beyond the flaming flowers ripples on vernal brook spreading;
Inside the white-washed walls willow and osmanthus blooming.

留园活泼泼地

Huo Po Po Di (The Place of Liveliness)
in Liu Yuan (The Lingering Garden)

綠徑穿花

Grass-paved path threading
through fields of flowers.

留园涵碧山房前铺地
Patterned pavement in front of Han Bi Shan Fang
(The Han Bi [Be imbued with the green] Mountain Villa) in
Liu Yuan (The Lingering Garden)

一庭芳草 疏风淡月有时来

The garden of flowers and grass, sweet and fair,
Graced now and then by pale moonlight and gentle air.

留园新建陈列馆

New built Exhibition Hall in
Liu Yuan (The Lingering Garden)

春在梨花院

Spring-sprinkled pear park.

狮子林立雪堂望院中
Courtyard of Li Xue Tang
(The Standing-in-Snow Hall)
in Shi Zi Lin (The Lion Forest Garden)

蘭堂有酒留佳客

To wine and dine guests at the
scented hall of orchid.

獅子林立雪堂望燕譽堂

Yan Yu Tang (The Hall of Peace and Happiness)
in Shi Zi Lin (The Lion Forest Garden), viewed from
Li Xue Tang (The Standing-in-Snow Hall)

昼永瑣愨閒　日遥簾幕静

During the day the sun shines on the shut window,
Toward the eve the screen still hangs quiet and low.

狮子林小方厅后院

Backyard of Xiao Fang Ting
(The Small Square Hall) in Shi Zi Lin
(The Lion Forest Garden)

Twilight befalls the quiet garden,
Setting peonies blazing with abandon.

狮子林牡丹台

A raised peony flower bed in
Shi Zi Lin (The Lion Forest Garden)

In search of serene beauty.

狮子林探幽门

Tan You Men (Places of Repose Sought)
in Shi Zi Lin (The Lion Forest Garden)

Bamboo pavilion excels all in the tranquil garden;
Seductive scene thrills one to high heaven.

曲亭深院 美景良辰

狮子林修竹阁

Xiu Zhu Ge (The Bamboo Pavilion)
in Shi Zi Lin (The Lion Forest Garden)

春绝何家来 试向溪边问

Whence the spring?
Hark to the stream murmuring!

狮子林扇面亭

Shan Mian Ting (The Fan Pavilion)
in Shi Zi Lin (The Lion Forest Garden)

The fig withered and the moss worn;
Fondle the steep rocks, if you care,
Only to find myriad peaks still shone with the light of moon.

狮子林指柏轩前

Front of Zhi Bo Xuan
(The Pointing at Cypress Hall)
in Shi Zi Lin (The Lion Forest Garden)

修水溅春　新篠溪绿　翠光交映虚亭

The stream tumbling at the height of spring,
Over the verdant meadow burgeoning sprigs swing,
And on the airy bower magic lights fling.

狮子林湖心亭

Hu Xin Ting (The Mid-Lake Pavilion)
in Shi Zi Lin (The Lion Forest Garden)

人去月侵廊 花下凌波入夢

People gone, silvery moonlight creeps into the porch;
Night down, flowers and ripples into the dreamland launch.

狮子林曲廊

A winding verandah in
Shi Zi Lin (The Lion Forest Garden)

新晴庭户春阴薄

The mild sun of spring sprinkles its light around;
The pale shade of pines sprawls the yard ground.

狮子林古五松园
Gu Wu Song Yuan
(The Garden of Five Age-old Pines)
in Shi Zi Lin (The Lion Forest Garden)

Mirror underwater are hills, ridge upon ridge,
Baffling the eye with their upturned images, edge after edge.

狮子林假山

Rockeries in Shi Zi Lin
(The Lion Forest Garden)

天峰飛墮地 翠岩誰削

Crashed aground from Heaven a bluish mount;
Whose divine hands hacked it out?

狮子林假山
Rockeries in Shi Zi Lin
(The Lion Forest Garden)

青松瘦竹　白石蒼苔

The sturdy pines snuggling by the slim and supple bamboo;
The gray rocks cloaked with moss, green or blue.

狮子林假山

Rockeries in Shi Zi Lin
(The Lion Forest Garden)

一泓澄綠 兩峽斬巖

A strip of water, crystal and green,
Flanked by rocks, steep and stern.

狮子林石洞
A cavern in Shi Zi Lin
(The Lion Forest Garden)

田田多少 一溪新綠漲晴波

Field after field stretching far and wide,
A stream of fresh green ripples flowing astride.

艺圃石桥
A stone bridge in Yi Pu
(The Garden of Cultivation)

Tricking water wearing away
the stepping stones.

艺圃小桥

A pedestrian stone bridge in
Yi Pu (The Garden of Cultivation)

池閣長春曉　園林綠暗紅稀

Over the pond, around the bower and flower bed,
Spring morn lingers with light spread,
Darkening the green and paling the red.

环秀山庄一角
A scenery spot in Huan Xiu Shan Zhuang
(The Mountain Villa with Embracing Beauty)

The tall trees toy with the sunset glow;
Along the dim dale swirling winds blow.

环秀山庄假山
Rockeries in Huan Xiu Shan Zhuang
(The Mountain Villa with Embracing Beauty)

小花深院静　日烘芳甃下罗藤

Delicate flowers in the deep and quiet yard grow;
Through wispy mist sun rays warm the vine below.

网师园小景

A scenery spot in
Wang Shi Yuan (The Master-of-nets Garden)

喚個月兒來 清光更多 只教冰壺一色

Sending for the Moon Fairy to shed her pearly light;
To turn the garden into a pond of silvery white.

网师园小石桥

A small pedestrian arch bridge in
Wang Shi Yuan (The Master-of-nets Garden)

(即"引静桥",
This bridge is called "The Leading to Quietude Bridge" now.)

玉阑春住 一帘芳景燕同吟

Spring delights in her stay within the jade rails;
Swallows sing over a maze of flowery trails.

网师园东部一角

A scenery spot in the eastern portion of
Wang Shi Yuan (The Master-of-nets Garden)

曲阑干外天如水

Beyond the winding rails, far into the horizon;
Heaven in water, water in heaven.

网师园长廊水亭

A Pavilion over the water along with roofed
walkway in Wang Shi Yuan (The Master-of-nets Garden)

(即"月到风来亭",
This pavilion is called "The Moon Comes With Breeze Pavilion" now.)

流水花間　曲榭芳亭細境

With water flowing and flowers blooming all around;
The charm of newly swept bower is even more profound.

网师园水亭

A Pavilion over the water in
Wang Shi Yuan (The Master-of-nets Garden)

(即"月到风来亭",
This pavilion is called "The Moon Comes With Breeze Pavilion" now.)

一鏡無塵　細草靜搖春碧

There lies a crystal-clear and jade-like pond;
With wispy grass in spring breeze swaying around.

网师园西北部

North west part of Wang Shi Yuan
(The Master-of-nets Garden)

花影透綺窗

A window embroidered with flower shadow.

怡园复廊漏窗

Lattice window in the double-corridor of
Yi Yuan (The Garden of Pleasance)

紫苔蒼壁 小園曲徑疏籬

Patches of greenish moss clings to the face of a grey crag;
The trail of a yard winds along a hedge loose and zigzag.

怡园拜石轩前

Front of Bai Shi Xuan
(Rock Fetish Pavilion) in Yi Yuan
(The Garden of Pleasance)

憑闌看落日　斷虹斜界雨初晴

I watch by the rail the rain-washed sunset flow,
And thrill at the sky-striding broken rainbow.

怡园藕香榭

Ou Xiang Xie (Lotus Scent Studio)
in Yi Yuan (The Garden of Pleasance)

小小新荷 點破清光景趣多

Unfolding their dainty and fresh green;
The newly-sprung lotuses enliven the scene.

怡园藕香榭前

Front of Ou Xiang Xie
(Lotus Scent Studio) in Yi Yuan
(The Garden of Pleasance)

Human shadows shuffling in the radiance of blossoms.

怡园镜中小沧浪

Xiao Cang Lang (The Small Cang Lang) in
Yi Yuan (The Garden of Pleasance), reflected in a mirror.

玉宇洗清秋　古木蒼烟無限意

Shower-cleansed, the fall air feels fresh and nice;
Mist-engulfed, the aged trees evoke emotional sighs.

怡园旱船

The Boat-Like Pavilion in
Yi Yuan (The Garden of Pleasance)

新篁嫩 摇碧玉 芳径绿苔生

In the fresh bamboo grove, tender green sprigs all sway;
On the grass-hugged path, verdant moss sprawls all the way.

怡园岁寒草庐前
Front of Sui Han Cao Lu
(Evergreen Thatched Cottage) in Yi Yuan
(The Garden of Pleasance)

Lake rocks of Yi Yuan.

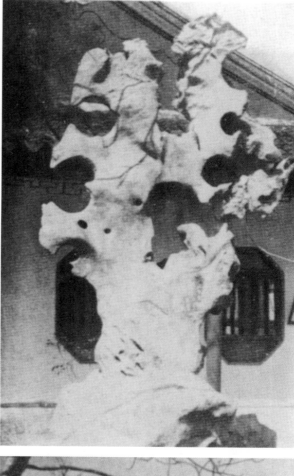

怡园湖石

A limestone in Yi Yuan
(The Garden of Pleasance)

Rugged rocks ranged round the window.

怡园湖石

A limestone in Yi Yuan
(The Garden of Pleasance)

杏花疏影裏 画帘低捲燕归来

Through the florid drapes up-rolled,
A flock of swallows are found back home,
In the dappled shade of apricot grove.

西白塔子巷李氏园

Li Shi Yuan (Li's Family Garden)
in Bai Ta Zi Xiang (White Pagota Alley)

滿身花影倩人扶

The shadow of flowers flashed on the wall,
Seems like the strokes of a fair girl.

西白塔子巷李氏園

Li Shi Yuan (Li's Family Garden)
in Bai Ta Zi Xiang (White Pagota Alley)

庭闲人静

Hushed Yard.

西白塔子巷李氏园

Li Shi Yuan (Li's Family Garden)
in Bai Ta Zi Xiang (White Pagota Alley)

Tranquil world beyond the moon gate.

西白塔子巷李氏园

Li Shi Yuan (Li's Family Garden)
in Bai Ta Zi Xiang (White Pagoda Alley)

花暗水房春

Spring graces the blossom-shaded waterside bower.

芳草园一角
Fang Cao Yuan (Fragrant Grass Garden)

柳上斜陽紅萬縷 烘人滿院荷香

The twilight turns the tops of willow into red radiance;
The lotus of the garden sends off whiffs of fragrance.

景德路慕家花园一角
Mu's Family Garden in Jing De Road

峭壁重霄 奇峰戳玉

The steep cliffs scrape massive clouds;
The jade peaks cap the mounts proud.

带城桥江苏师范附属女中瑞云峰
Rui Yun Feng (The Auspicious Cloud Peak)
located in the Girl's Secondary School associated to
Jiangshu Normal College at Dai Cheng Qiao

燕語鶯嬌小院幽 春色不分休

Swallows are chirping;
Orioles are singing.
On the quiet and greenish yard,
Those messengers offspring are joyfully calling.

韩家巷鹤园一角

He Yuan (The Crane Garden)
at Han Jia Xiang (Han's Family Alley)

新枝媚斜日 小园自有四时花

Bewitched by the spring, the twilight on willowy branches showers;
Called by the nature, the yard yields all seasonal flowers.

醋库巷柴氏园

Chai Shi Yuan (The Chai's Garden)
at Cu Ku Xiang

冷石生雲

Cool sweaty rocks.

铁瓶巷顾宅花厅
Hua Ting (Flower Hall)
in Gu's House at Tie Ping Xiang

梅风池渚　虹雨苔滋

As the plum blooms, the pond overflows;
Under the rainbow, the moss grows.

惠荫园紫藤

Chinese Wisteria in Hui Yin Yuan

乔木莺初啭　深院燕交飞

Orioles twitter tunefully in the arbor brush;
Swallows shuttle buoyantly in the yard without rush.

西百花巷程氏园

Cheng's Garden in Xi Bai Hua Xiang

幽径破苔痕　雕玉阑干深院静

Through the green moss wades a trodden trail;
Around the serene yard winds a carved rail.

西百花巷程氏园花厅

Hua Ting (Flower Hall) in
Cheng's Garden in Xi Bai Hua Xiang

红日长　绿烟晴　流莺三两声

The sun shines bright and long;
The jade-green leaves glisten off and on;
A darting oriole chants a liquid song.

沧浪亭前可园一角

A corner of Ke Yuan (The Passable Garden)
in front of the Garden of Cang Lang Ting
(The Cang Lang Pavilion)

古林疏又密　日烘芳炷下藤萝

Sparse here and dense there the age-old trees grow;
Through the coiling mist the sun blazes the vine below.

藕园花厅前一角

A corner in the front of *Hua Ting*
(Flower Hall) in *Ou Yuan* (The Couple's Garden Retreat)

臨池飛閣乍青紅

A tall lotus hall stands over a pond;
With green leaves and red flowers thronged.

藕园东部荷花厅

He Hua Ting (Lotus hall)
in the east part of Ou Yuan
(The Couple's Garden Retreat)

满庭岩桂谧香风

Sweet scent of osmanthus suffusing the yard.

藕园西部藏书楼

The Library Tower in the west part
of Ou Yuan (The Couple's Garden Retreat)

Full-swing spring flooding the idle.

小院開窗春已深

藕园西部
West part of Ou Yuan
(The Couple's Garden Retreat)

Into the house wafts floral fragrance of the night,
With bracing breeze and bright moonlight.
Isn't it the best time for delight.

曲房花氣窗 清風明月好時光

藕园西部
West part of Ou Yuan
(The Couple's Garden Retreat)

Myriad flowers frolic on the railed stage;
Muted birds delight in the dense flower cage.

藕园西部曲廊

A winding verandah in the west part
of Ou Yuan (The Couple's Garden Retreat)

人静静 日迟迟 深院断无人到

Not any sound,
Daytime lingers around,
In the yard no rambler is found.

藕园西部小斋

A small house in the west part
of Ou Yuan (The Couple's Garden Retreat)

林霏散浮暝　水光月色兩相累

Fog lifts in the woods at night;
On the lake the moon sheds its pale light.

戒幢寺西园水漏窗

Lattice window over the water
in Jie Zhuang Shi (The West Garden)

曲檻俯清流 晴空漾翠浪

Winding rail stoops over watery pearls;
Fresh breeze stirs up emerald swirls.

戒幢寺西园湖心亭

Hu Xin Ting (The Mid-lake Pavilion)
in Jie Zhuang Shi (The West Garden)

墙角移阴

The sun in the sky strides;
The shadow in the corner hides.

景德路杨氏园

Yang's Garden in Jing De Road

芭蕉籠碧砌　秋聲先到簾櫳

Palms trees thrive by the marble-girded meadow;
Autumn air breathes into curtained window.

沧浪亭

Cang Lang Ting (The Cang Lang Pavilion)

花影壁重門

Flower-hugged gate.

景德路杨氏园园门

Entrance Gate of Yang's Garden in Jing De Road

The green-groomed bower reveling
in late autumn glow.

一樓擎翠生秋暝

沧浪亭漏窗

Lattice window in the Garden of
Cang Lang Ting (The Cang Lang Pavilion)

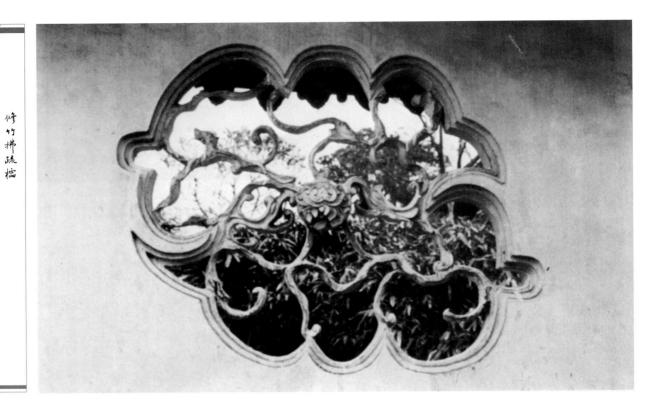

修竹拂疏櫺

Dainty bamboo fondling the airy lattice.

沧浪亭漏窗

Lattice window in the Garden of
Cang Lang Ting (The Cang Lang Pavilion)

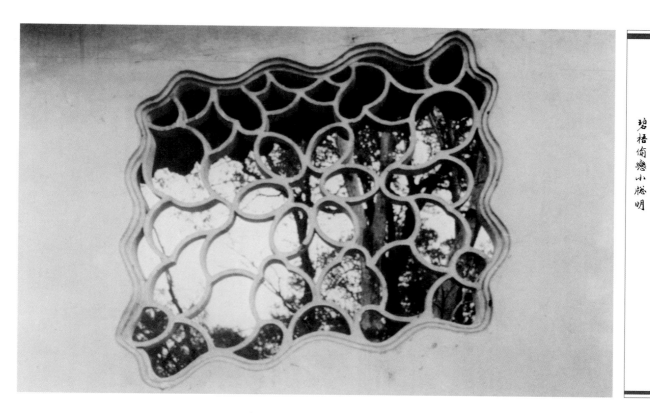

The phoenix tree coveting the
sun-kissed window.

沧浪亭漏窗

Lattice window in the Garden of
Cang Lang Ting (The Cang Lang Pavilion)

翠陰初轉午間花 深院聽啼鶯

Verdant shades revolve around mid-day blossom;
Merry orioles twitter in the yard's bosom.

沧浪亭御碑亭

Yu Bei Ting (The Imperial Stele Pavilion) in the
Garden of Cang Lang Ting (The Cang Lang Pavilion)

水閣無塵午晝長

Water-washed pavilion basked in splendid sunshine.

沧浪亭复廊

The double-corridor in the Garden
of Cang Lang Ting (The Cang Lang Pavilion)

曲径通深窈 有墙头桂影 窗上梅花

Winding pathway extends far, far away;
Shadows of osmanthus on the wall sway;
Plum blossoms by the window merrily play.

沧浪亭步碕亭廊

Corridor in Bu Qi Ting (The Bu Qi Pavilion)
in the Garden of Cang Lang Ting (The Cang Lang Pavilion)

古木通幽徑 翠微站窒 晚烟半隱平林

Through ancient trees winds a tranquil trail;
Emerald luster shimmers all the way to a blue hill;
A grove is dimly visible in the haze of a dusk dale.

沧浪亭亭前

In front of the Cang Lang Ting
(The Cang Lang Pavilion) in the Garden of
Cang Lang Ting (The Cang Lang Pavilion)

濃綠漲瑤窗 亂紅迷紫曲

Verdant sprigs by the jade window grow;
Red and violet are all lost in a dazzling glow.

仓米巷半园三曲桥

Three fold zigzag bridge in Ban Yuan
(Half Garden) located in Cang Mi Xiang

綠楊庭院 青杏園林 月影又移牆影

In the yard poplars are everywhere found;
Around the garden apricot trees abound;
On the wall moonlight nudges shadows around.

畅园西面一角

West side of Chang Yuan
(The Garden of Free Will)

占断雕阑只一枝 春风费尽几工夫

For a branch to outgrow the carved balustrade,
Alas, how much efforts has vernal breeze made.

畅园西面一角
West side of Chang Yuan
(The Garden of Free Will)

樓臺還清峭　冷螢隨月到迴廊

A pavilion stands quiet on a crisp and starry autumn night;
Along the corridor crickets sing under the frosty moonlight.

畅园东面一角

East side of Chang Yuan
(The Garden of Free Will)

深院鎖春風 尘蓬綠池幽 交枝徑窄

The deep garden traps vernal breeze;
The grey clouds hover over the pond like green fleece;
The slim path is laced with entangled trees.

畅园曲廊月门

Winding verandah and Moon gate
in Chang Yuan (The Garden of Free Will)

芳径雨初晴

Rain-flushed and flower-scented trail.

畅园小径铺地

Patterned pavement on a trail
in Chang Yuan (The Garden of Free Will)

水縱橫 石蔥蘢

Crisscrossed waterways;
Elegantly-wrought rockery.

附：无锡寄畅园石桥

Annex: A stone bridge in Ji Chang Yuan
(Garden for Lodging One's Expansive Feelings)
in the city of Wuxi

紅粉晴隨流水去　園林漸覺清陰密

Red petals are washed quietly down the rill,
Leaving the garden clearer and greener still.

附：无锡寄畅园

Annex: Ji Chang Yuan
(Garden for Lodging One's Expansive Feelings)
in the city of Wuxi

幽懷倚石

Reclining against a rock in contemplation.

附：无锡梅园

Annex: Mei Yuan (Plum Garden)
in the city of Wuxi

苏州拙政园测绘图录

Architectural Survey of
Zhuo Zheng Yuan
(The Humble Administrator's Garden)
in Suzhou

拙政园平面图
Site Plan of Zhuo Zheng Yuan (The Humble Administrator's Garden), scale unit in meter

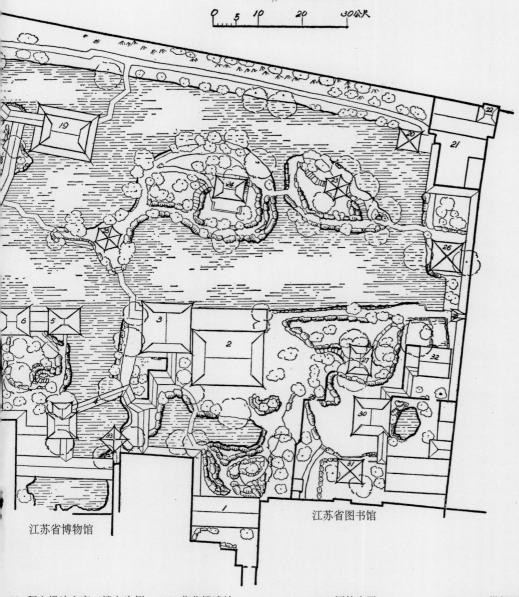

江苏省博物馆　　江苏省图书馆

18. 拜文揖沈之斋（楼上为倒影楼）
18. Bai Wen Yi Shen Zhi Zhai (The Room for Worshipping Wen and Shen, ground floor of Dao Ying Lou (The Tower of Reflection)
19. 见山楼
19. Jian Shan Lou (The Mountain-In-View Tower)
20. 绿漪亭
20. Lu Yi Ting (The Green Ripple Pavilion)
21. 菜花楼遗址
21. Ruin of Cai Hua Lou
22. 料敌楼
22. Liao Di Lou
23. 待霜亭
23. Dai Shuang Ting (The Orange Pavilion)
24. 雪香云蔚亭
24 Xue Xiang Yun Wei Ting (The Snow-Like Fragrant Prunus Mume Pavilion)
25. 荷风四面亭
25. He Feng Si Mian Ting (The Pavilion in Lotus Breezes)
26. 梧竹幽居
26. Wu Zhu You Ju (The Secluded Pavilion of Firmiana Simplex and Bamboo)
27. 绣绮亭
27. Xiu Qi Ting (The Paeonia Suffruticosa Pavilion)
28. 东半亭（倚虹亭）
28. Dong Ban Ting (East Half Pavilion) or Yi Hong Ting (Leaning against Rainbow Pavilion)
29. 枇杷园入口
29. Entrance of Pi Pa Yuan (Loquat Garden Court)
30. 玲珑馆
30. Ling Long Guan (The Hall of Elegance)
31. 嘉实亭
31. Jia Shi Ting (The Loquat Pavilion)
32. 海棠春坞
32. Hai Tang Chun Wu (The Malus Spectabilis Garden Court)
33. 松风亭
33. Song Feng Ting

上：拙政园中部横断面图
Upper: Cross section of the middle part of Zhu

下：拙政园远香堂倚玉轩小飞虹及香洲
Lower: Yuan Xiang Tang (The Hall of Distant
Fei Hong (The Small Flying Rainbow Bridge) a
Yuan (The Humble Administrator's Garden)

Zheng Yuan (The Humble Administrator's Garden)

agrance), Yi Yu Xuan (The Bamboo Pavilion), Xiao
Xiang Zhou (The Fragrant Isle) in Zhuo Zheng

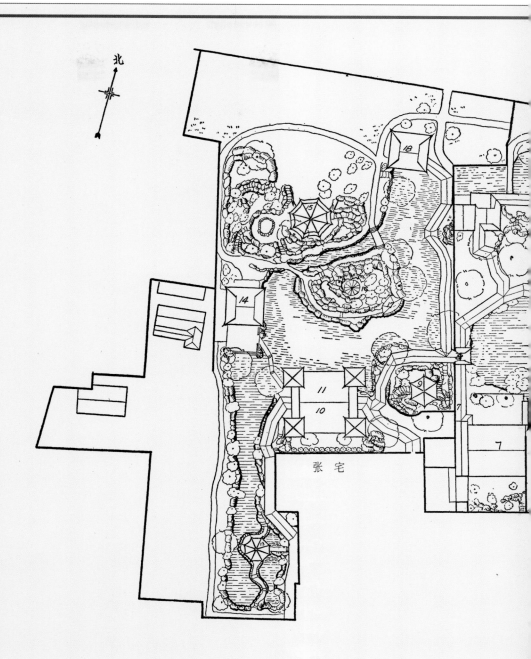

1. 腰门
1. Yao Men
2. 远香堂
2. Yuan Xiang Tang (The Hall of Distant Fragrance)
3. 倚玉轩
3. Yi Yu Xuan (The Bamboo Pavilion)
4. 小飞虹
4. Xiao Fei Hong (The Small Flying Rainbow Bridge)
5. 香洲
5. Xiang Zhou (The Fragrant Isle)
6. 澂观楼
6. Cheng Guang Lou
7. 玉兰堂
7. Yu Lan Tang (The Magnolia Hall)
8. 别有洞天
8. Bie You Dong Tian (Another Cave World)
9. 宜两亭
9. Yi Liang Ting (The Good for Both Families Pavilion)
10. 十八曼陀罗花馆
10. Shi Ba Man Tuo Luo Hua Guan (The Hall of 18 Camellias)
11. 三十六鸳鸯馆
11. San Shi Liu Yuan Yang Guan (The Hall of Thirty-six Pairs of Mandarin Ducks)
12. 云坞
12. Yun Wu
13. 塔影亭
13. Ta Ying Ting (The Pagoda Reflection Pavilion)
14. 留听阁
14. Liu Ting Ge (The Stay and Listen Pavilion)
15. 浮翠阁
15. Fu Cui Ge (The Floating Green Tower)
16. 笠亭
16. Li Ting (The Indus Calamus Pavilion)
17. 与谁同坐轩（扇亭）
17. Yu Shei Tong Zuo Xuan ("With Whom Shall I Sit?" Pavilion), or Shan Ting (The Fan Pavilion).

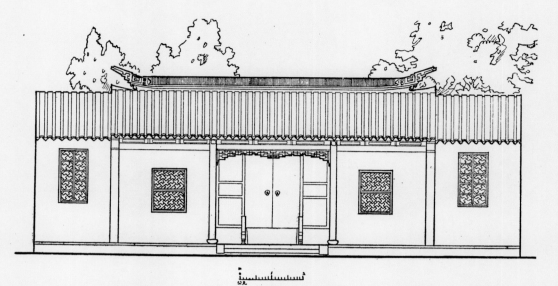

腰门正立面图
Front elevation of Yao Men

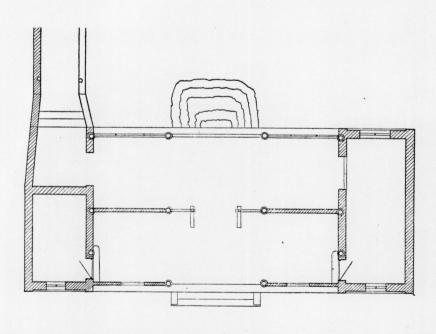

腰门平面图
Floor plan of Yao Men

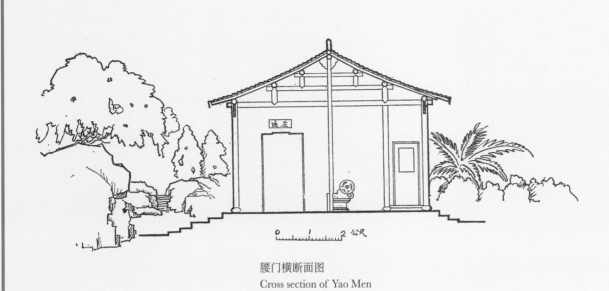

腰门横断面图
Cross section of Yao Men

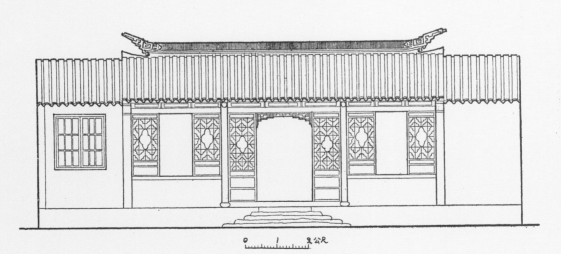

腰门背立面图
Back elevation of Yao Men

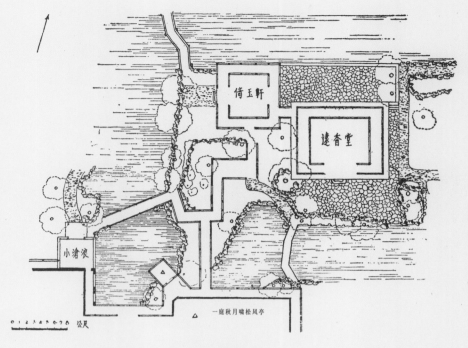

远香堂环境图
Site Plan of Yuan Xiang Tang
(The Hall of Distant Fragrance)

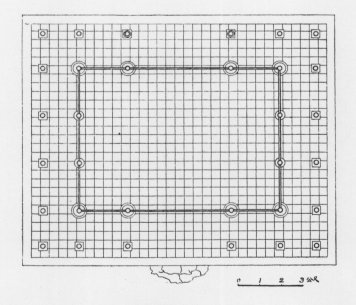

远香堂平面图
Floor plan of Yuan Xiang Tang
(The Hall of Distant Fragrance)

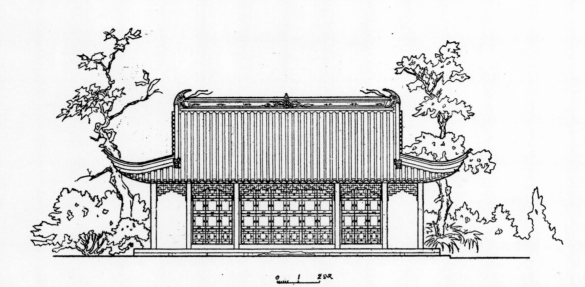

远香堂正立面图
Front elevation of Yuan Xiang Tang
(The Hall of Distant Fragrance)

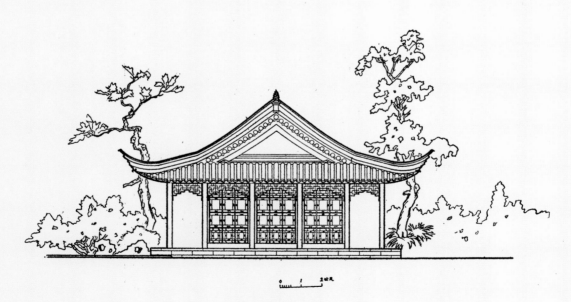

远香堂侧立面图
Side elevation of Yuan Xiang Tang
(The Hall of Distant Fragrance)

倚玉轩正立面图
Front elevation of Yi Yu Xuan
(The Bamboo Pavilion)

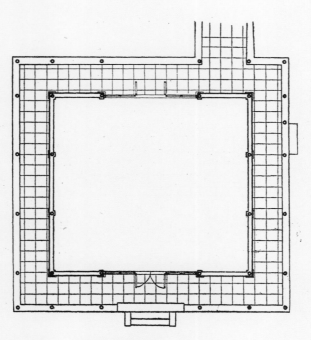

倚玉轩平面图
Floor plan of Yi Yu Xuan
(The Bamboo Pavilion)

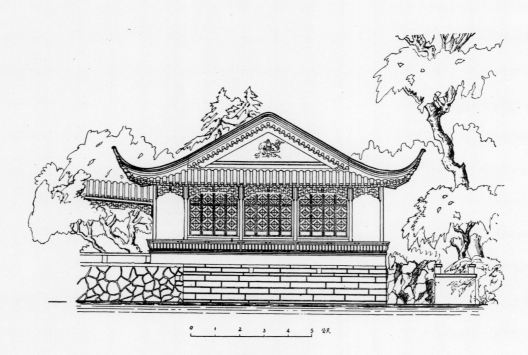

倚玉轩侧立面图
Side elevation of Yi Yu Xuan
(The Bamboo Pavilion)

绣绮亭平面图
Floor plan of Xiu Qi Ting
(The Paeonia Suffruticosa Pavilion)

绣绮亭立面图
Elevation of Xiu Qi Ting
(The Paeonia Suffruticosa Pavilion)

绣绮亭横断面图
Cross section of Xiu Qi Ting
(The Paeonia Suffruticosa Pavilion)

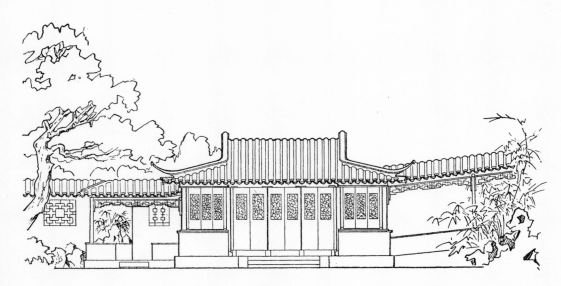

玲珑馆正立面图
Front elevation of Ling Long Guan
(The Hall of Elegance)

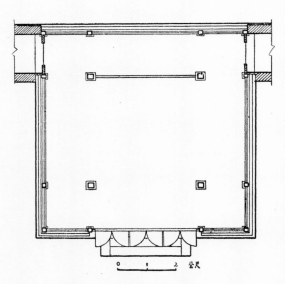

玲珑馆平面图
Floor plan of Ling Long Guan
(The Hall of Elegance)

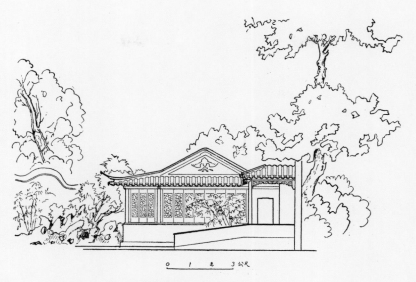

玲珑馆侧立面图
Side elevation of Ling Long Guan
(The Hall of Elegance)

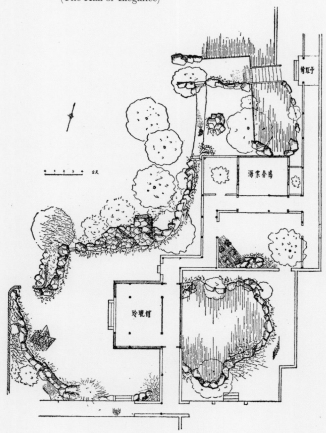

玲珑馆及海棠春坞环境图
Site plan of Ling Long Guan (The Hall of Elegance)
and Hai Tang Chun Wu (The Malus Spectabilis Garden Court)

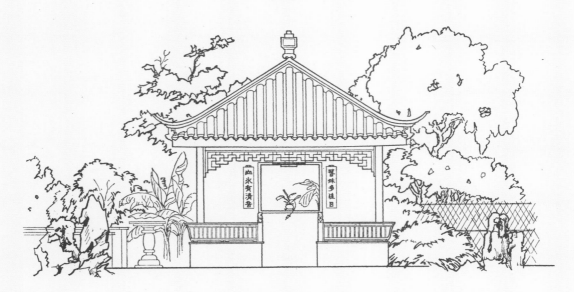

嘉实亭正立面图
Front elevation of Jia Shi Ting
(The Loquat Pavilion)

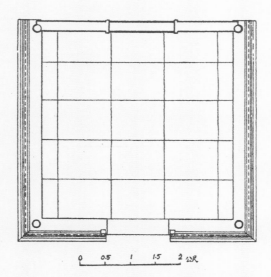

嘉实亭平面图
Floor plan of Jia Shi Ting
(The Loquat Pavilion)

倚虹亭正立面图
Front elevation of Yi Hong Ting
(Leaning against Rainbow Pavilion)

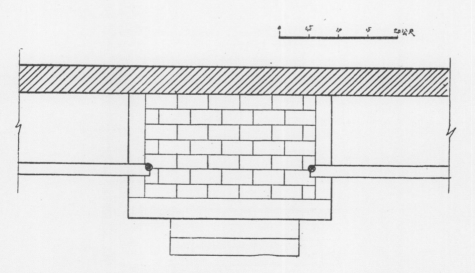

倚虹亭平面图
Floor plan of Yi Hong Ting
(Leaning against Rainbow Pavilion)

海棠春坞正立面图
Front elevation of Hai Tang Chun Wu
(The Malus Spectabilis Garden Court)

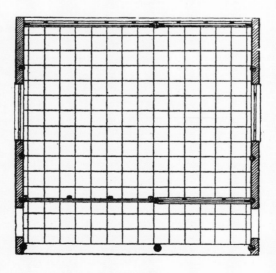

海棠春坞平面图
Floor plan of Hai Tang Chun Wu
(The Malus Spectabilis Garden Court)

梧竹幽居正立面图
Front elevation of Wu Zhu You Ju
(The Secluded Pavilion of Firmiana Simplex and Bamboo)

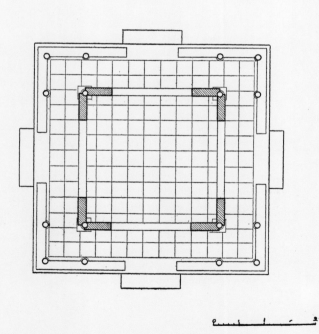

梧竹幽居平面图
Floor plan of Wu Zhu You Ju (The Secluded
Pavilion of Firmiana Simplex and Bamboo)

绿漪亭正立面图
Front elevation of Lu Yi Ting
(The Green Ripple Pavilion)

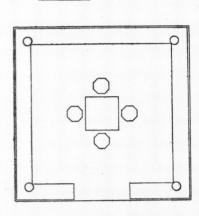

绿漪亭平面图
Floor plan of Lu Yi Ting (The Green Ripple Pavilion)

待霜亭正立面图
Front elevation of Dai Shuang Ting
(The Orange Pavilion)

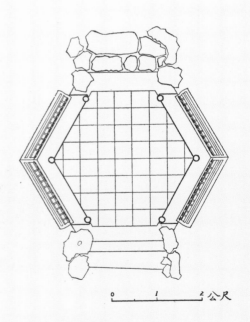

待霜亭平面图
Floor plan of Dai Shuang Ting
(The Orange Pavilion)

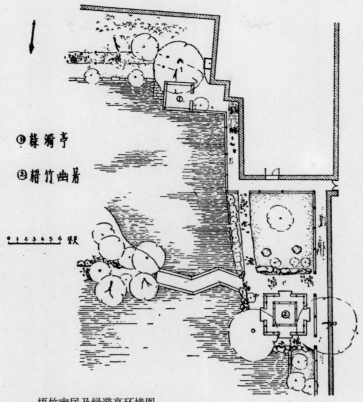

梧竹幽居及绿漪亭环境图
Site plan of Hai Tang Chun Wu (The Malus Spectabilis Garden Court) and Lu Yi Ting (The Green Ripple Pavilion)

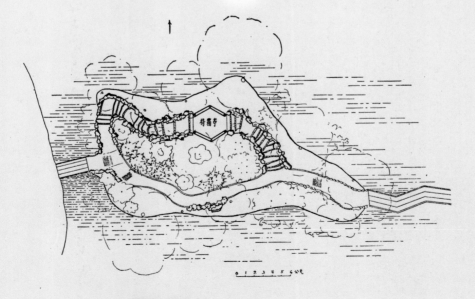

待霜亭环境图
Site plan of Dai Shuang Ting (The Orange Pavilion)

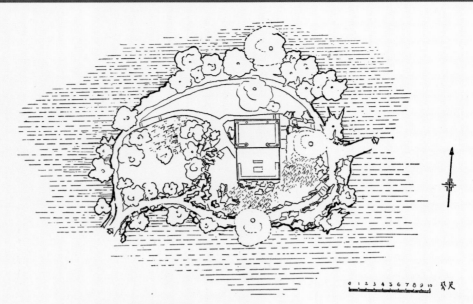

雪香云蔚亭环境图
Site plan of Xue Xiang Yun Wei Ting (The Snow-Like Fragrant Prunus Mume Pavilion)

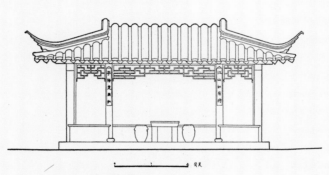

雪香云蔚亭正立面图
Front elevation of Xue Xiang Yun Wei Ting
(The Snow-Like Fragrant Prunus Mume Pavilion)

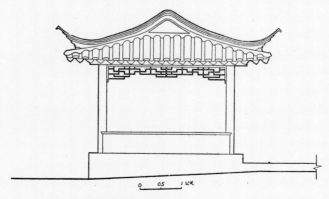

雪香云蔚亭侧立面图
Side elevation of Xue Xiang Yun Wei Ting
(The Snow-Like Fragrant Prunus Mume Pavilion)

雪香云蔚亭正面图
Front elevation of Xue Xiang Yun Wei Ting
(The Snow-Like Fragrant Prunus Mume Pavilion)

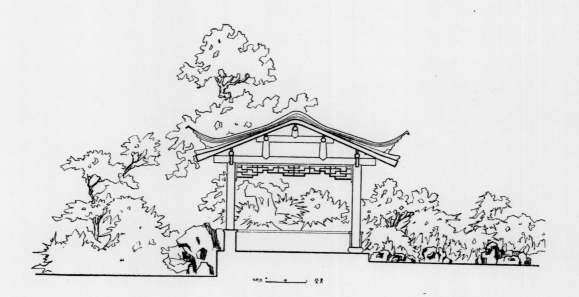

雪香云蔚亭横断面图
Cross Section of Xue Xiang Yun Wei Ting
(The Snow-Like Fragrant Prunus Mume Pavilion)

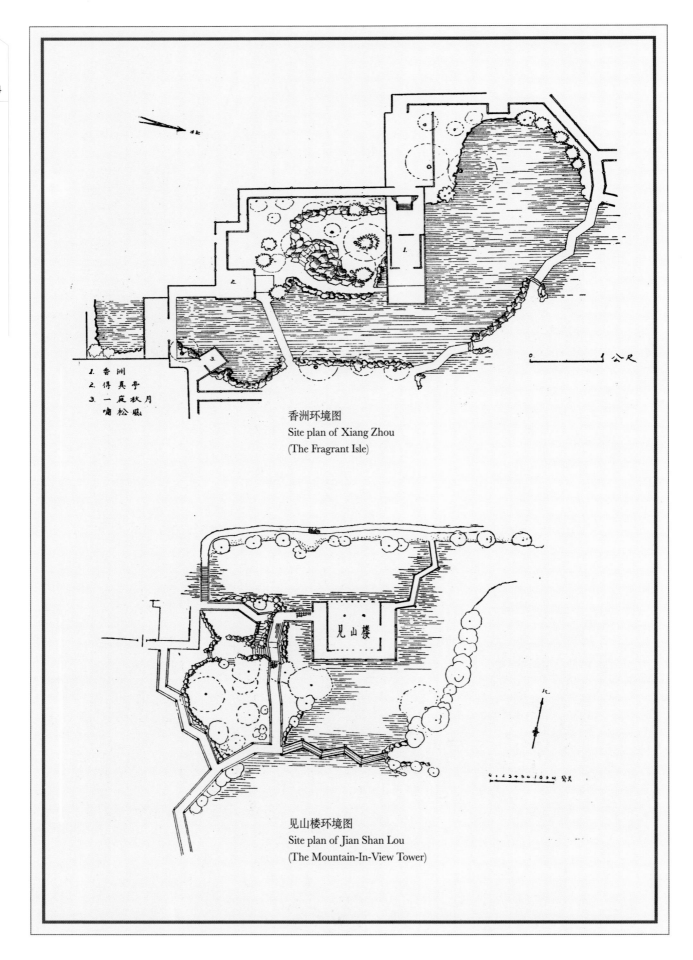

香洲环境图
Site plan of Xiang Zhou
(The Fragrant Isle)

见山楼环境图
Site plan of Jian Shan Lou
(The Mountain-In-View Tower)

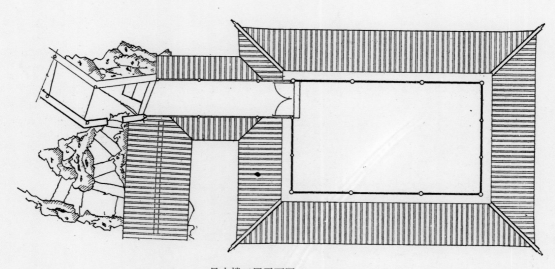

见山楼二层平面图
Floor plan of the second floor of Jian Shan Lou
(The Mountain-In-View Tower)

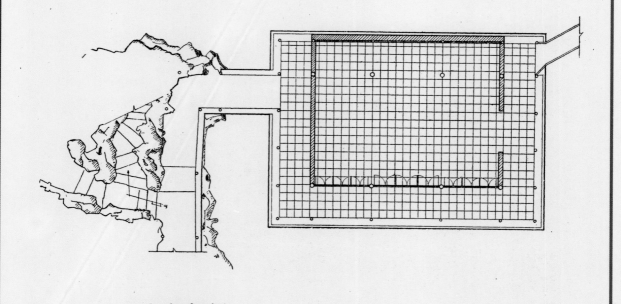

见山楼底层平面图
Floor plan of the ground floor of Jian Shan Lou
(The Mountain-In-View Tower)

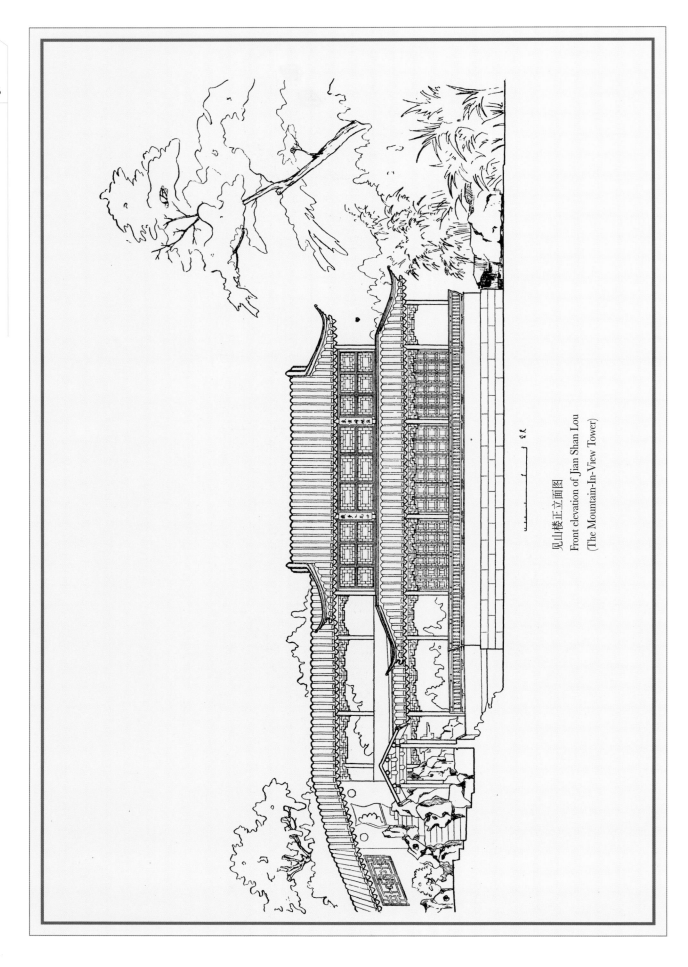

见山楼正立面图
Front elevation of Jian Shan Lou
(The Mountain-In-View Tower)

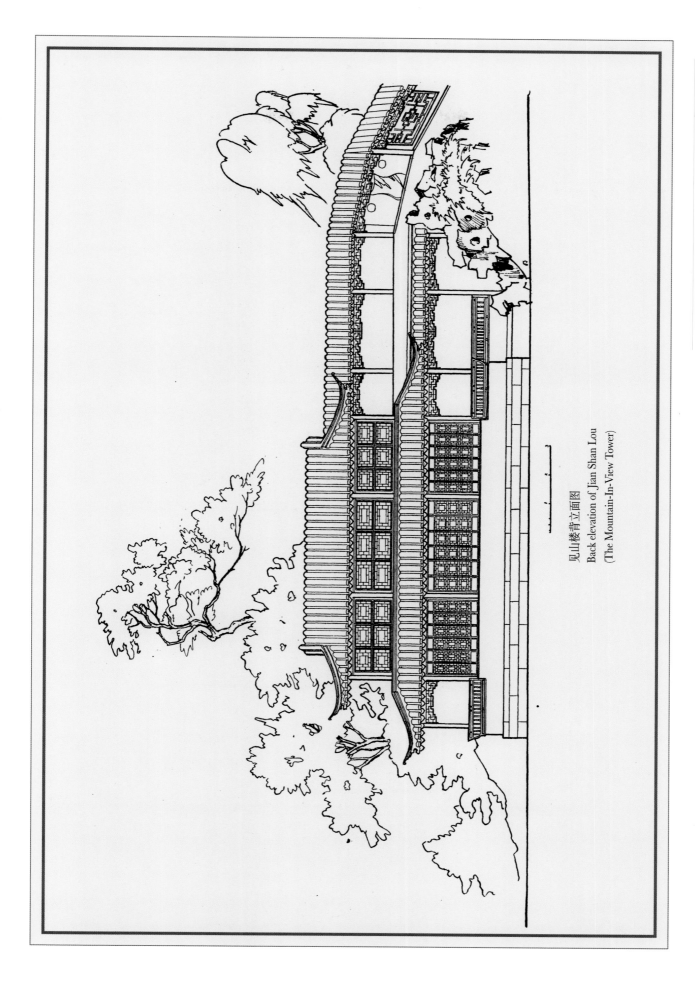

见山楼背立面图
Back elevation of Jian Shan Lou
(The Mountain-In-View Tower)

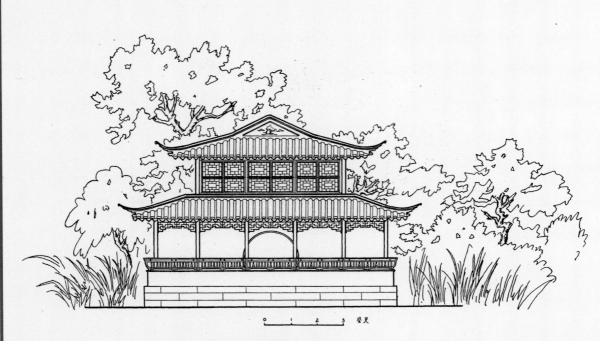

见山楼侧立面图
Side elevation of Jian Shan Lou
(The Mountain-In-View Tower)

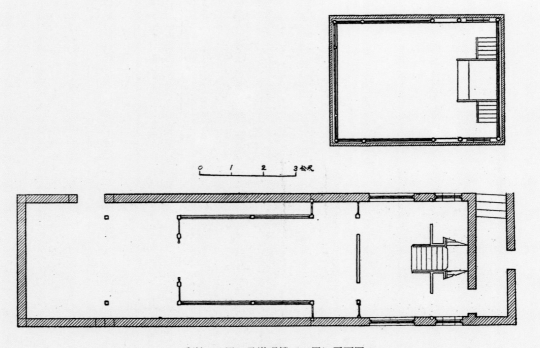

香洲（一层）及澄观楼（二层）平面图
Floor plans of Xiang Zhou (The Fragrant Isle, first floor) and Cheng Guang Lou (second floor)

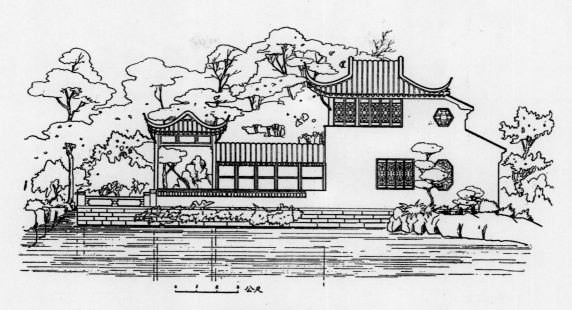

香洲及澄观楼侧立面图
Side elevation of Xiang Zhou (The Fragrant Isle) and Cheng Guang Lou

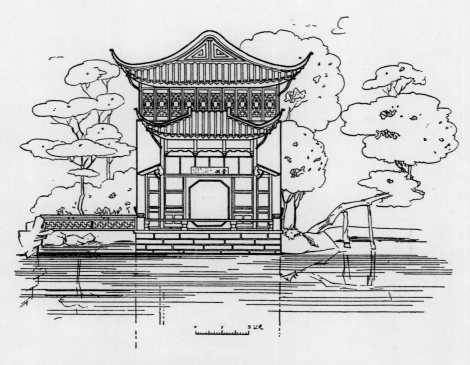

香洲及澄观楼正立面图
Front elevation of Xiang Zhou (The Fragrant Isle) and Cheng Guang Lou

得真亭正立面图
Front elevation of De Zhen Ting
(The True Nature Pavilion)

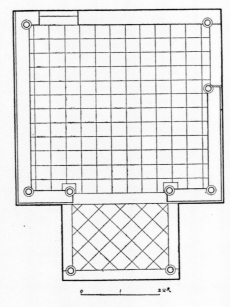

得真亭平面图
Floor plan of De Zhen Ting
(The True Nature Pavilion)

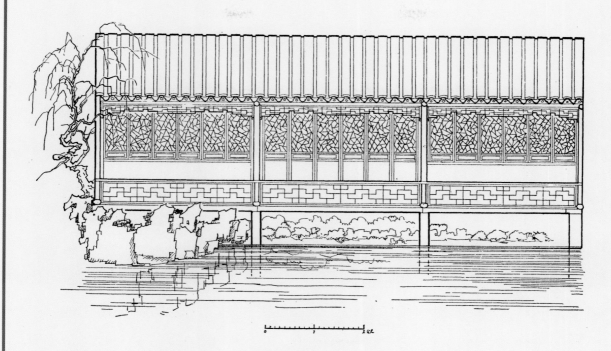

小沧浪正立面图
Front elevation of Xiao Cang Lang
(The Small Cang Lang)

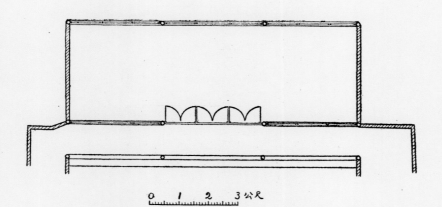

小沧浪平面图
Floor plan of Xiao Cang Lang
(The Small Cang Lang)

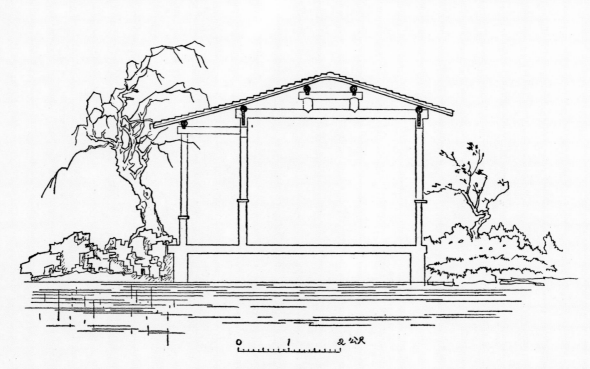

小沧浪横断面图
Cross Section of Xiao Cang Lang
(The Small Cang Lang)

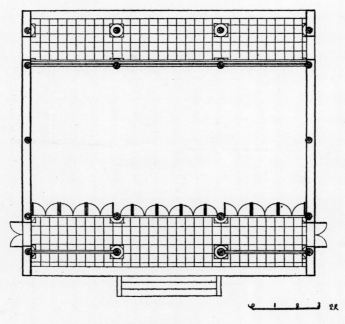

玉兰堂平面图
Floor plan of Yu Lan Tang
(The Magnolia Hall)

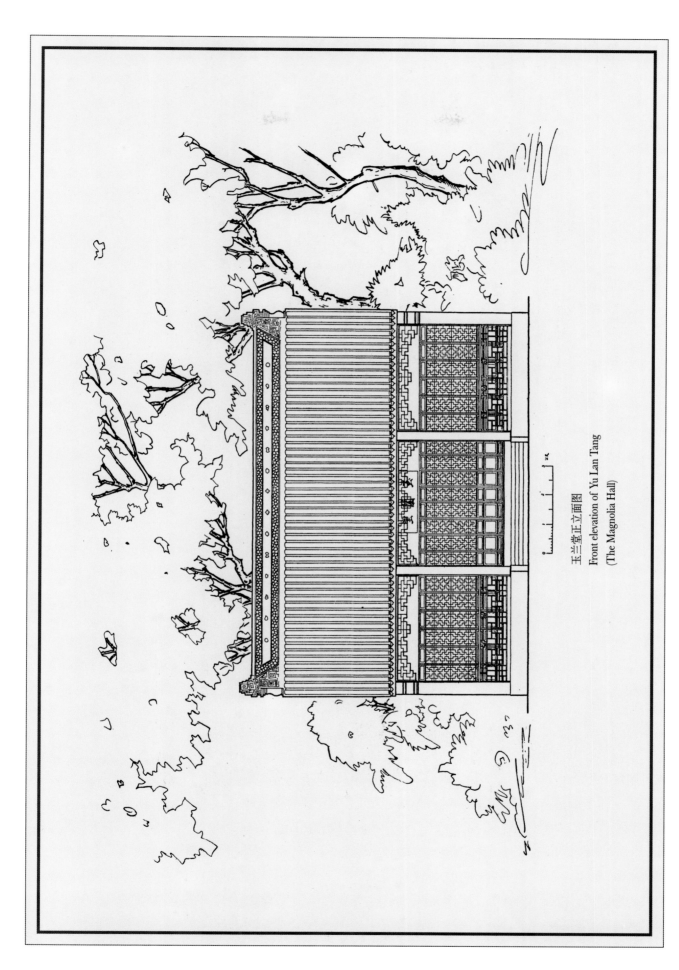

玉兰堂正立面图
Front elevation of Yu Lan Tang
(The Magnolia Hall)

一庭秋月啸松风亭立面图
Elevation of Yi Ting Qiu Yue Xiao Song Feng Ting
(Autumn Moon and Windy Pine Pavilion)

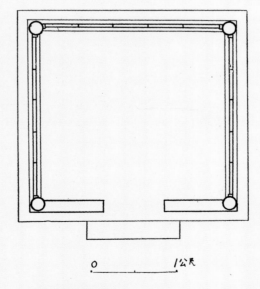

一庭秋月啸松风亭平面图
Floor plan of Yi Ting Qiu Yue Xiao Song Feng Ting
(Autumn Moon and Windy Pine Pavilion)

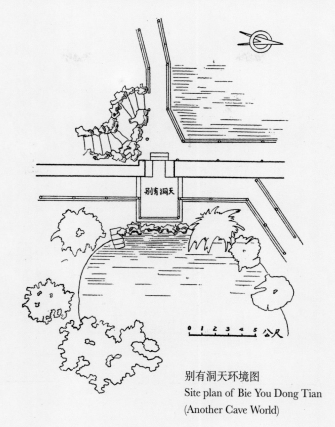

别有洞天环境图
Site plan of Bie You Dong Tian
(Another Cave World)

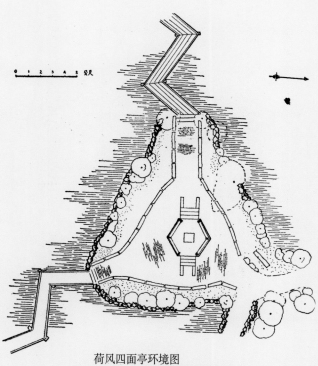

荷风四面亭环境图
Site plan of He Feng Si Mian Ting
(The Pavilion in Lotus Breezes)

荷风四面亭正立面图
Front elevation of He Feng Si Mian Ting
(The Pavilion in Lotus Breezes)

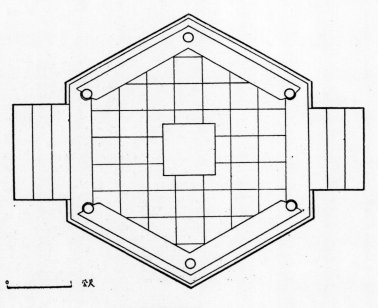

荷风四面亭平面图
Floor plan of He Feng Si Mian Ting
(The Pavilion in Lotus Breezes)

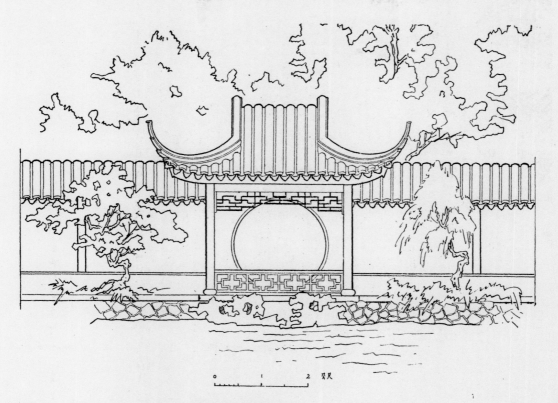

别有洞天正立面图
Front elevation of Bie You Dong Tian
(Another Cave World)

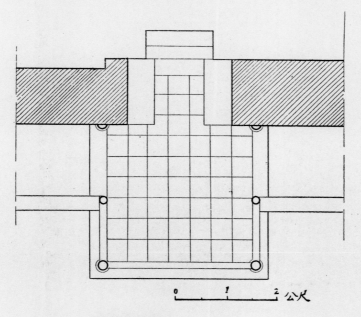

别有洞天平面图
Floor plan of Bie You Dong Tian
(Another Cave World)

笠亭立面图
Elevation of Li Ting
(The Indus Calamus Pavilion)

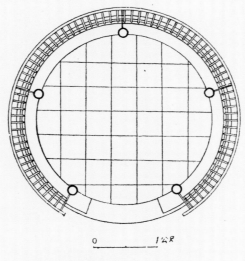

笠亭平面图
Floor plan of Li Ting
(The Indus Calamus Pavilion)

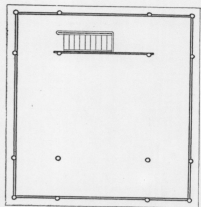

倒影楼（二层）平面图
Second floor plan of Dao Ying Lou
(The Tower of Reflection)

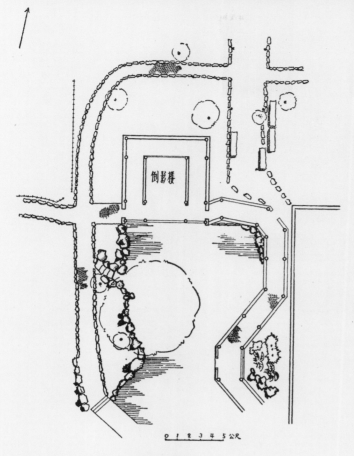

倒影楼环境图
Site plan of Dao Ying Lou
(The Tower of Reflection)

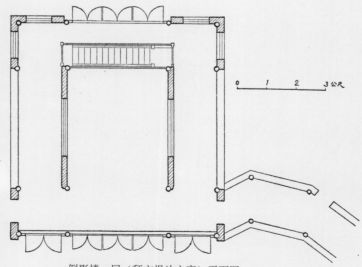

倒影楼一层（拜文揖沈之斋）平面图
Ground floor plan of Dao Ying Lou
(The Room for Worshipping Wen and Shen)

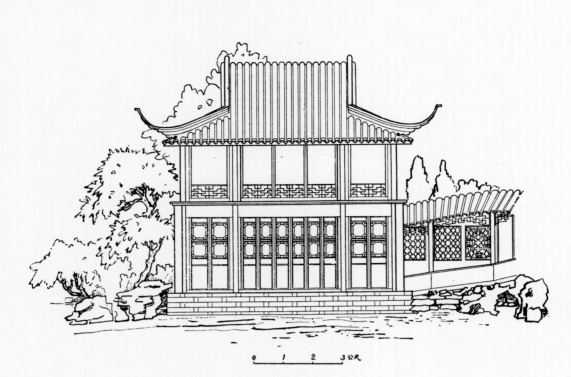

倒影楼正立面图
Front elevation of Dao Ying Lou
(The Tower of Reflection)

倒影楼侧立面图
Side elevation of Dao Ying Lou
(The Tower of Reflection)

宜两亭立面图
Front elevation of Yi Liang Ting
(The Good for Both Families Pavilion)

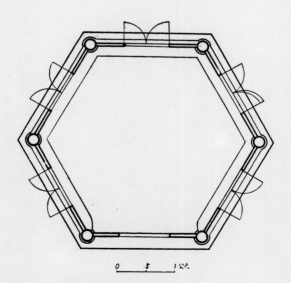

宜两亭平面图
Floor plan of Yi Liang Ting
(The Good for Both Families Pavilion)

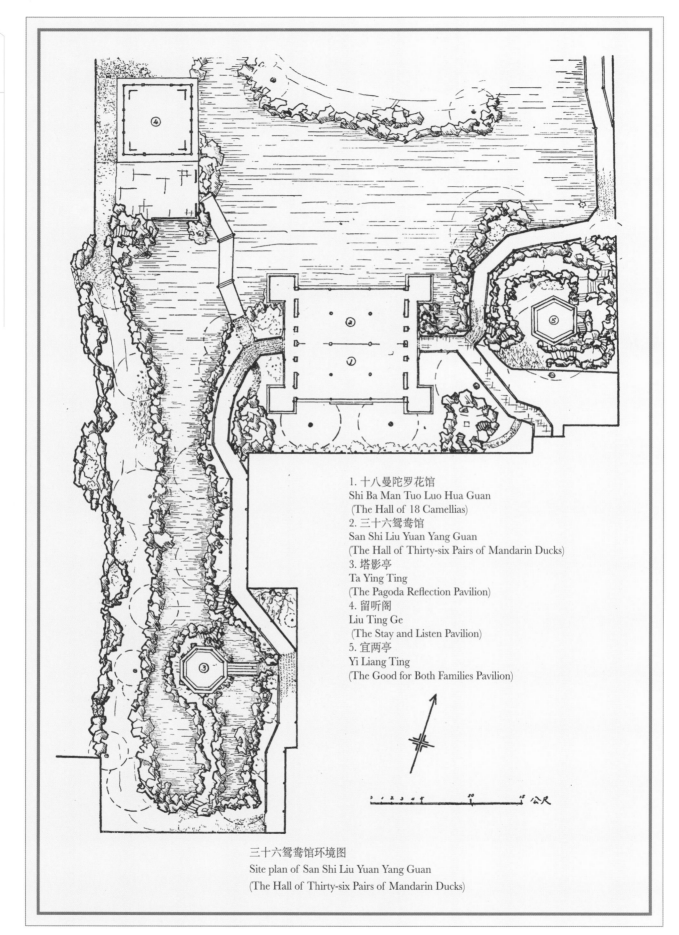

1. 十八曼陀罗花馆
Shi Ba Man Tuo Luo Hua Guan
(The Hall of 18 Camellias)
2. 三十六鸳鸯馆
San Shi Liu Yuan Yang Guan
(The Hall of Thirty-six Pairs of Mandarin Ducks)
3. 塔影亭
Ta Ying Ting
(The Pagoda Reflection Pavilion)
4. 留听阁
Liu Ting Ge
(The Stay and Listen Pavilion)
5. 宜两亭
Yi Liang Ting
(The Good for Both Families Pavilion)

三十六鸳鸯馆环境图
Site plan of San Shi Liu Yuan Yang Guan
(The Hall of Thirty-six Pairs of Mandarin Ducks)

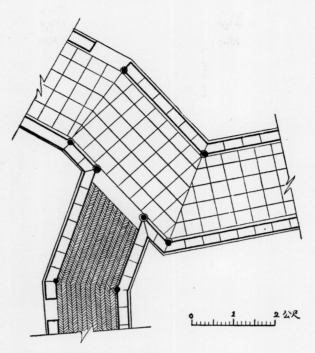

三十六鸳鸯馆曲廊平面图
Floor plan of winding verandah of San Shi Liu Yuan Yang Guan
(The Hall of Thirty-six Pairs of Mandarin Ducks)

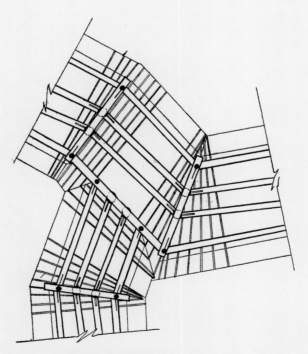

三十六鸳鸯馆曲廊屋顶仰视图
Ceiling plan of winding verandah of San Shi Liu Yuan Yang Guan
(The Hall of Thirty-six Pairs of Mandarin Ducks)

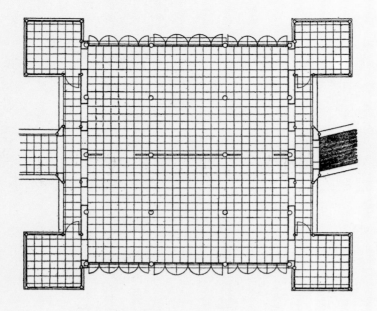

三十六鸳鸯馆及十八曼陀罗花馆平面图
Floor plan of San Shi Liu Yuan Yang Guan (The Hall of Thirty-six Pairs of Mandarin Ducks) and Shi Ba Man Tuo Luo Hua Guan (The Hall of 18 Camellias)

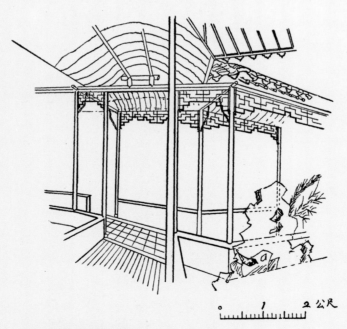

三十六鸳鸯馆曲廊透视图
Perspective view of winding verandah of San Shi Liu Yuan Yang Guan (The Hall of Thirty-six Pairs of Mandarin Ducks)

三十六鸳鸯馆临水立面图
Water side elevation of San Shi Liu Yuan Yang Guan
(The Hall of Thirty-six Pairs of Mandarin Ducks)

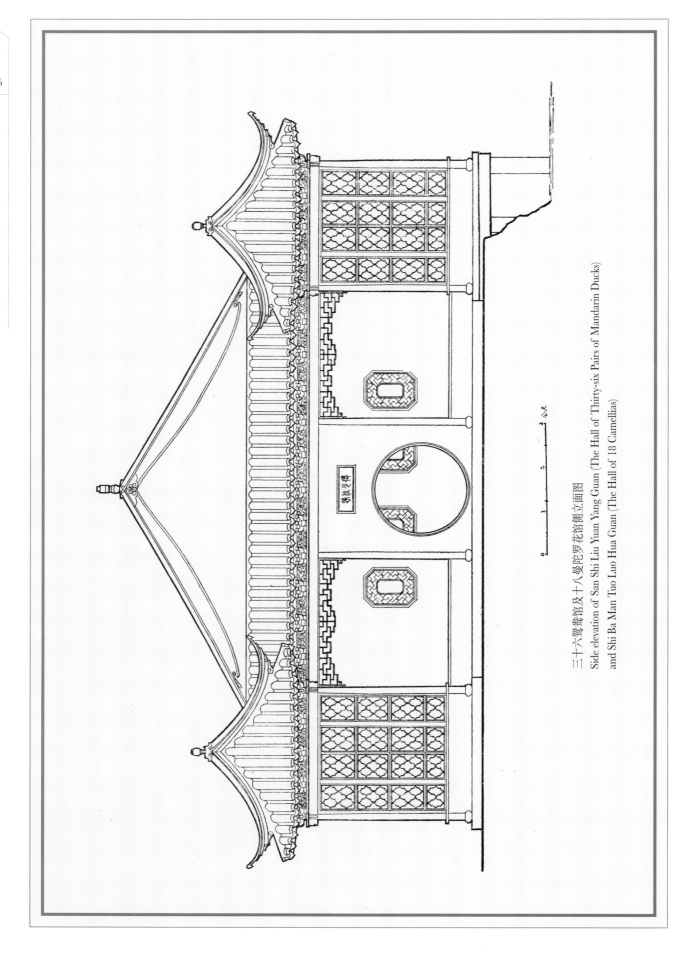

三十六鸳鸯馆及十八曼陀罗花馆侧立面图
Side elevation of San Shi Liu Yuan Yang Guan (The Hall of Thirty-six Pairs of Mandarin Ducks) and Shi Ba Man Tuo Luo Hua Guan (The Hall of 18 Camellias)

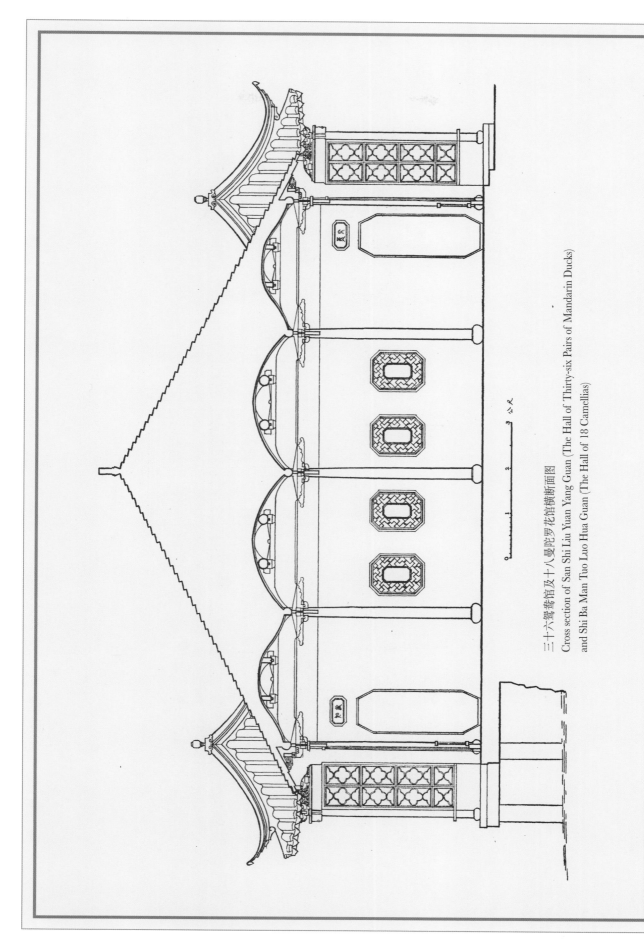

三十六鸳鸯馆及十八曼陀罗花馆横断面图
Cross section of San Shi Liu Yuan Yang Guan (The Hall of Thirty-six Pairs of Mandarin Ducks) and Shi Ba Man Tuo Luo Hua Guan (The Hall of 18 Camellias)

留听阁正立面图
Front elevation of Liu Ting Ge
(The Stay and Listen Pavilion)

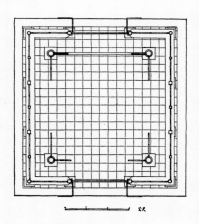

留听阁平面图
Floor plan of Liu Ting Ge
(The Stay and Listen Pavilion)

留听阁侧立面图
Side elevation of Liu Ting Ge
(The Stay and Listen Pavilion)

塔影亭立面图
Elevation of Ta Ying Ting
(The Pagoda Reflection Pavilion)

塔影亭俯视图
Top view of Ta Ying Ting
(The Pagoda Reflection Pavilion)

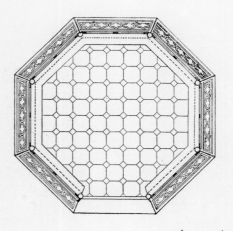

塔影亭平面图
Floor plan of Ta Ying Ting
(The Pagoda Reflection Pavilion)

与谁同坐轩正立面图
Front elevation of Yu Shei Tong Zuo Xuan
("With Whom Shall I Sit?" Pavilion)

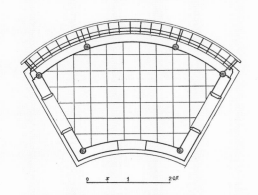

与谁同坐轩平面图
Floor plan of Yu Shei Tong Zuo Xuan
("With Whom Shall I Sit?" Pavilion)

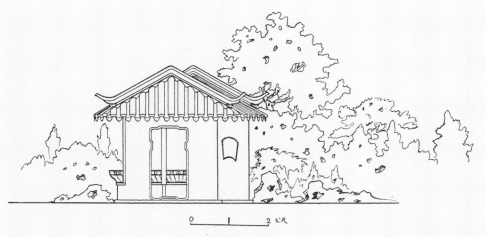

与谁同坐轩侧立面图
Side elevation of Yu Shei Tong Zuo Xuan
("With Whom Shall I Sit?" Pavilion)

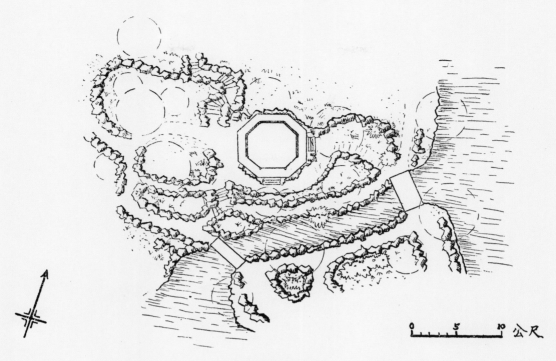

浮翠阁环境图
Site plan of Fu Cui Ge
(The Floating Green Tower)

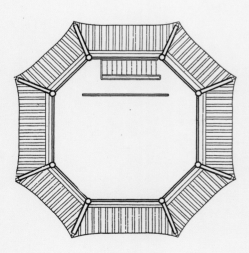

浮翠阁二层平面图
Floor plan of the second floor Fu Cui Ge
(The Floating Green Tower)

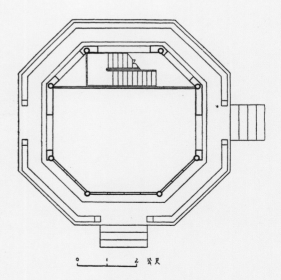

浮翠阁底层平面图
Floor plan of the ground floor Fu Cui Ge
(The Floating Green Tower)

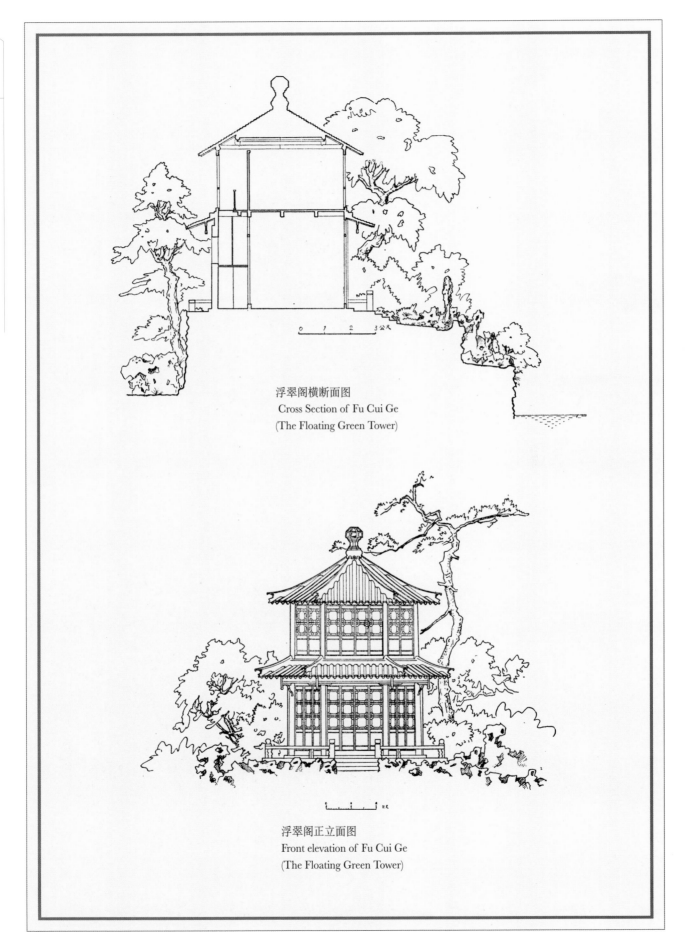

浮翠阁横断面图
Cross Section of Fu Cui Ge
(The Floating Green Tower)

浮翠阁正立面图
Front elevation of Fu Cui Ge
(The Floating Green Tower)

苏州留园测绘图录

◆

Architectural Survey of
Liu Yuan
(The Lingering Garden)
in Suzhou

留园平面图
Site plan of Liu Yuan (The Lingering Garden) in SuZhou

1. 门厅
1. MenTing
2. 古木交柯
2. Gu Mu Jiao Ke (The Intertwined Old Trees)
3. 绿荫
3. Lu Ying (The Green Shade Pavilion)
4. 涵碧山房
4. Han Bi Shan Fang (The Han Bi [Be imbued with the green] Mountain Villa)
5. 凉台
5. Liang Tai (Terrace at the north of Han Bi Shan Fang [The Han Bi (Be imbued with the green) Mountain Villa])
6. 明瑟楼
6. Ming Se Lou (The Pellucid or The Pure Freshness Tower)
7. 闻木樨香轩
7. Wen Mu Xi Xiang Xuan (The Osmanthus Fragrance Pavilion)

8. 可亭
8. Ke Ting (The Passable Pavilion)
9. 远翠阁
9. Yuan Cui Ge (The Distant Green Tower)
10. 小蓬莱
10. Xiao Peng Lai (The Small Fairy Isle)
11. 濠濮亭
11. Hao Pu Ting (The Hao Pu Pavilion)
12. 曲谿楼
12. Qu Xi Lou (The Winding Stream Tower)
13. 西楼
13. Xi Lou (Western Tower)
14. 清风池馆
14. Qing Feng Chi Guan (The Refreshing Breeze Pavilion by the Lake)
15. 五峰仙馆
15. Wu Feng Xian Guan (The Celestial Hall of Five Peaks)

16. 汲古得绠处
16. Ji Gu De Geng Chu (The Study of Enlightenment)
17. 还我读书处
17. Huan Wo Du Shu Chu (The Return-to-Read Study)
18. 揖峰轩
18. Yi Feng Xuan (The Worshipping Stone Pavilion)
19. 林泉耆硕之馆
19. Lin Quan Qi Shuo Zhi Guan (The Old Hermit Scholars' Hall)
20. 亦不二亭
20. Yi Bu Er Ting
21. 仁云庵
21. Zhu Yun An
22. 冠云楼
22. Guan Yun Lou (The Cloud-Capped Tower)
23. 冠云亭
23. Guan Yun Ting (The Cloud-Capped Pavilion)

24. 冠云台
24. Guan Yun Tai
25. 浣云沼
25. Huan Yun Zhao
26. 东园一角亭
26. A pavilion at the corner of eastern portion
27. 佳晴喜雨快雪之亭
27. Jia Qing Xi Yu Kuai Xue Zhi Ting (The Good-For-Farming Pavilion)
28. 陈列馆
28. Exhibition Hall
29. 至乐亭
29. Zhi Le Ting (The Delightful Pavilion)
30. 舒啸亭
30. Shu Xiao Ting (The Free Roaring Pavilion)
31. 活泼泼地
31. Huo Po Po Di (The Place of Liveliness)
32. 水阁
32. Shui Ge

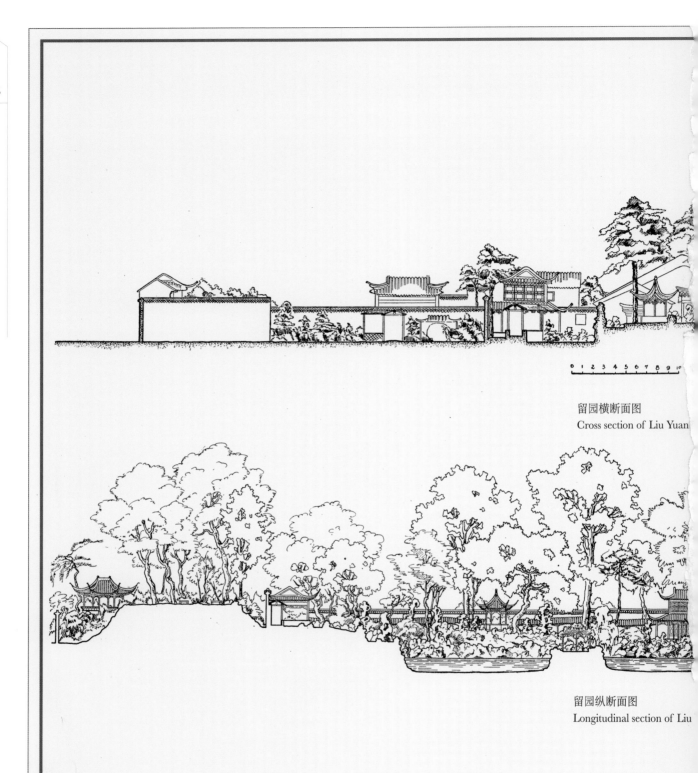

留园横断面图
Cross section of Liu Yuan

留园纵断面图
Longitudinal section of Liu

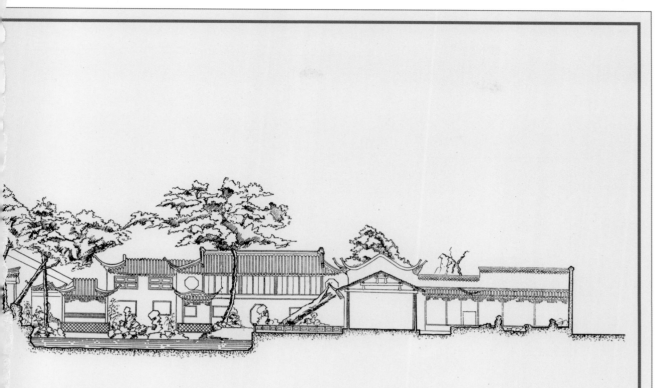

(The Lingering Garden)

Yuan (The Lingering Garden)

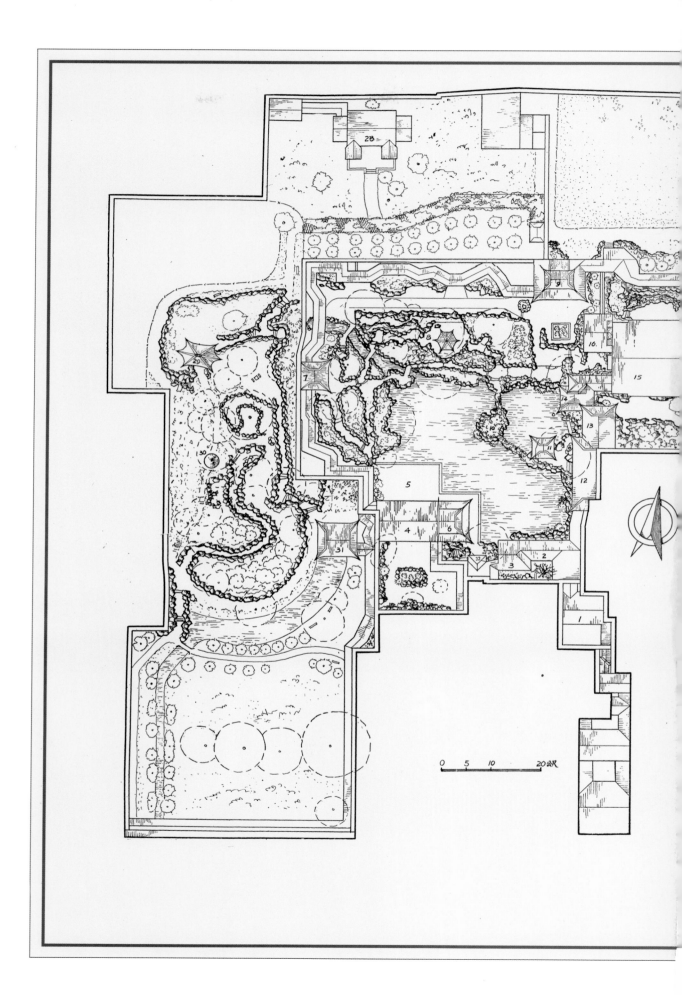

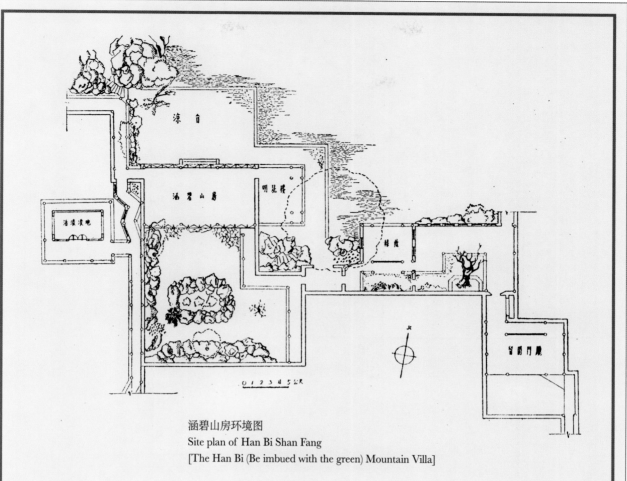

涵碧山房环境图
Site plan of Han Bi Shan Fang
[The Han Bi (Be imbued with the green) Mountain Villa]

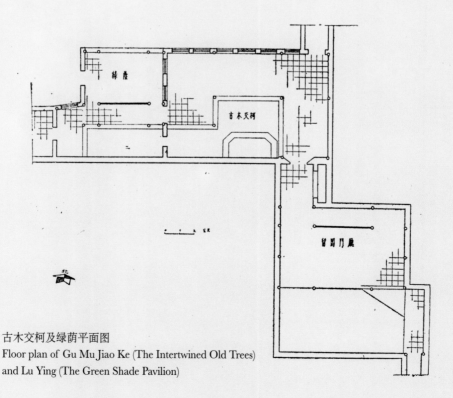

古木交柯及绿荫平面图
Floor plan of Gu Mu Jiao Ke (The Intertwined Old Trees)
and Lu Ying (The Green Shade Pavilion)

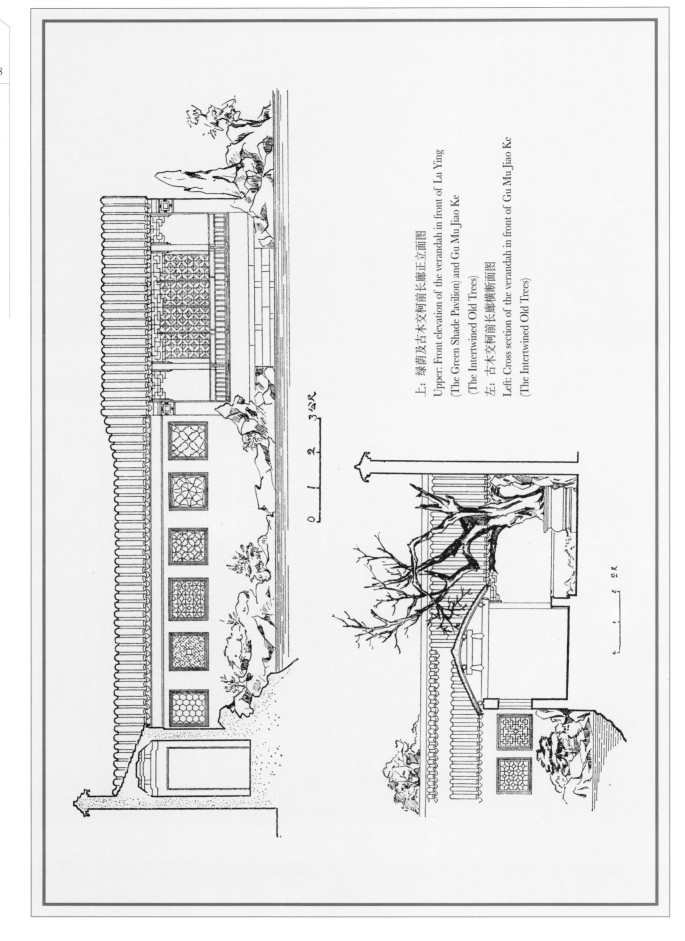

上：绿荫及古木交柯前长廊正立面图
Upper: Front elevation of the verandah in front of Lu Ying (The Green Shade Pavilion) and Gu Mu Jiao Ke (The Intertwined Old Trees)
左：古木交柯前长廊横断面图
Left: Cross section of the verandah in front of Gu Mu Jiao Ke (The Intertwined Old Trees)

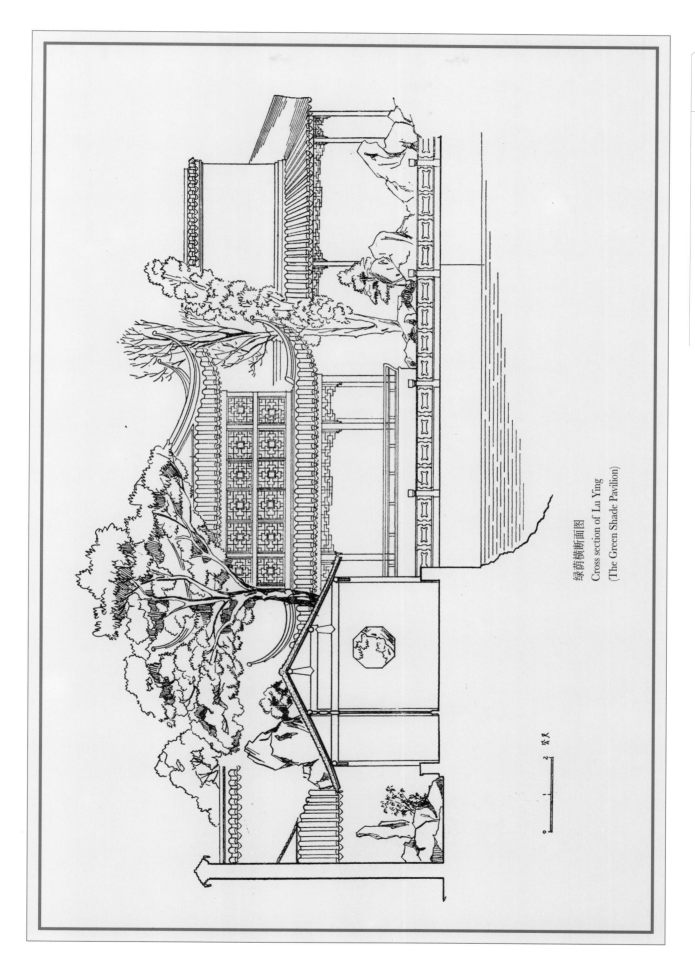

绿荫横断面图
Cross section of Lu Ying
(The Green Shade Pavilion)

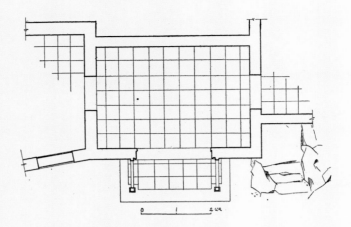

水阁平面图
Floor plan of Shui Ge

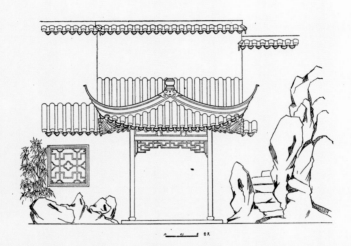

水阁正立面图
Front elevation of Shui Ge

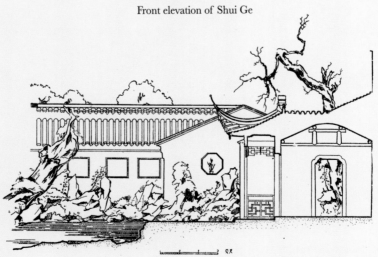

水阁横断面图
Cross section of Shui Ge

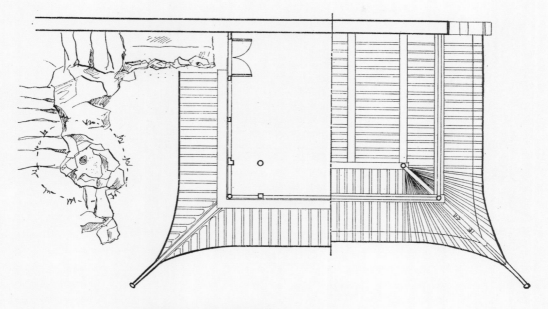

明瑟楼二层平面及屋顶仰视图
Floor plan and mirror reflection of ceiling plan, Second floor of Ming Se Lou (The Pellucid or The Pure Freshness Tower)

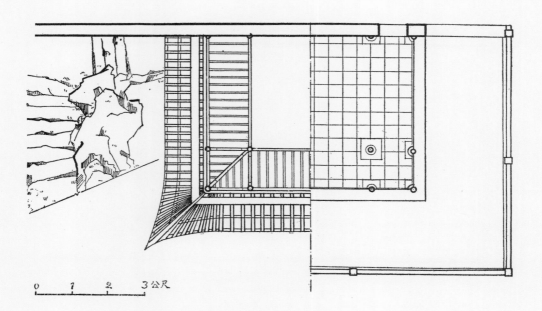

明瑟楼底层平面及上层仰视图
Floor plan and mirror reflection of ceiling plan, Ground floor of Ming Se Lou (The Pellucid or The Pure Freshness Tower)

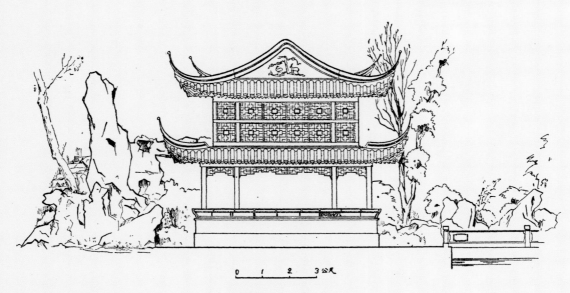

明瑟楼侧立面图
Side elevation of Ming Se Lou
(The Pellucid or The Pure Freshness Tower)

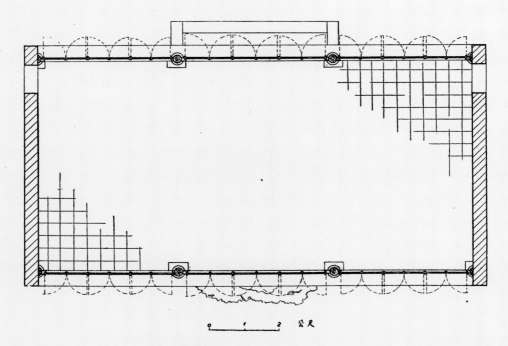

涵碧山房平面图
Floor plan of Han Bi Shan Fang
(The Han Bi (Be imbued with the green) Mountain Villa)

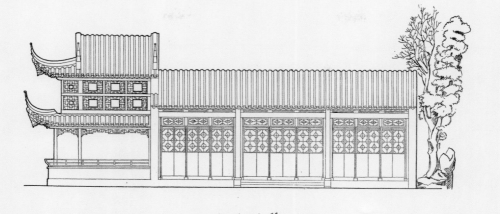

涵碧山房背立面图及明瑟楼正立面图
Back elevation of Han Bi Shan Fang (The Han Bi [Be imbued with the green] Mountain Villa) and front elevation of Ming Se Lou (The Pellucid or The Pure Freshness Tower)

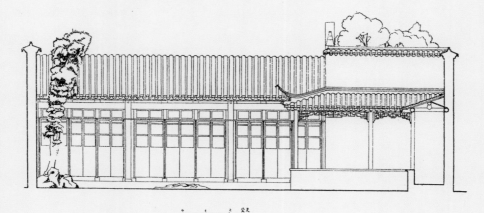

涵碧山房正立面图
Front elevation of Han Bi Shan Fang
(The Han Bi [Be imbued with the green] Mountain Villa)

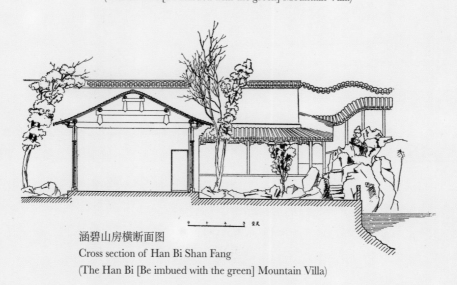

涵碧山房横断面图
Cross section of Han Bi Shan Fang
(The Han Bi [Be imbued with the green] Mountain Villa)

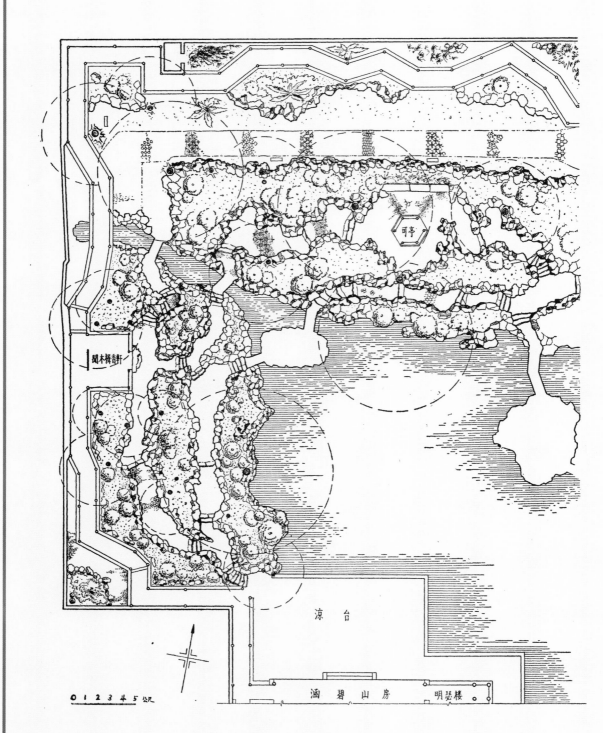

闻木樨香轩环境图
Site plan of Wen Mu Xi Xiang Xuan
(The Osmanthus Fragrance Pavilion)

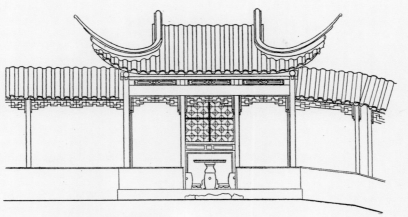

闻木樨香轩正立面图
Front elevation of Wen Mu Xi Xiang Xuan
(The Osmanthus Fragrance Pavilion)

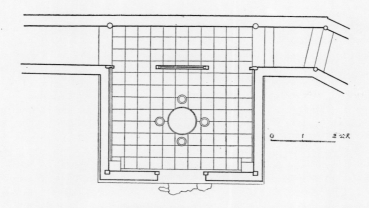

闻木樨香轩平面图
Floor plan of Wen Mu Xi Xiang Xuan
(The Osmanthus Fragrance Pavilion)

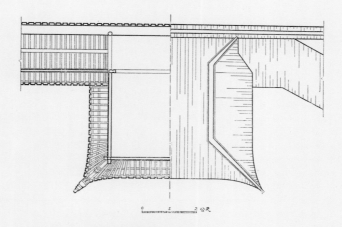

闻木樨香轩屋顶仰视及俯视图
Mirror reflection of ceiling plan and roof plan, Wen Mu Xi Xiang Xuan
(The Osmanthus Fragrance Pavilion)

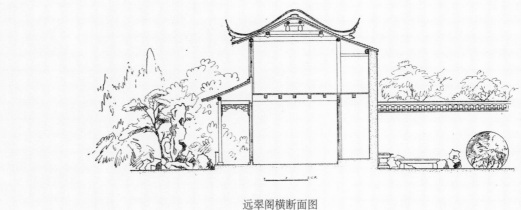

远翠阁横断面图
Cross section of Yuan Cui Ge
(The Distant Green Tower)

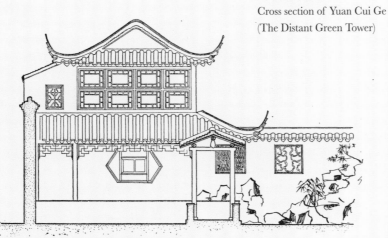

远翠阁侧立面图
Side elevation of Yuan Cui Ge
(The Distant Green Tower)

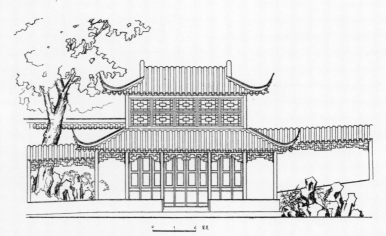

远翠阁正立面图
Front elevation of Yuan Cui Ge
(The Distant Green Tower)

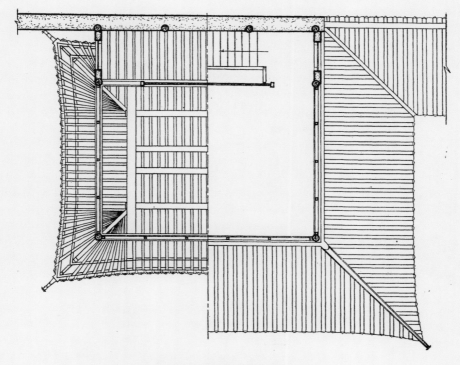

远翠阁二层平面及屋顶仰视图
Floor plan and mirror reflection of ceiling plan,
Second floor of Yuan Cui Ge (The Distant Green Tower)

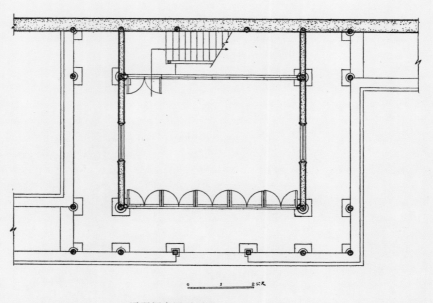

远翠阁底层平面图
Floor plan of Yuan Cui Ge
(The Distant Green Tower)

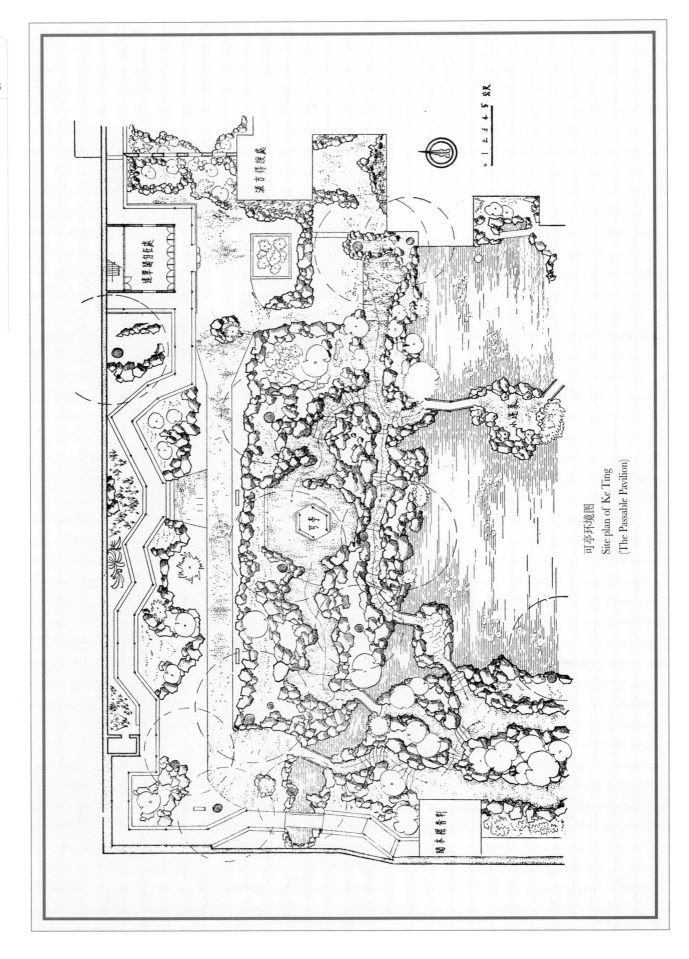

可亭环境图
Site plan of Ke Ting
(The Passable Pavilion)

可亭断面图
Section of Ke Ting
(The Passable Pavilion)

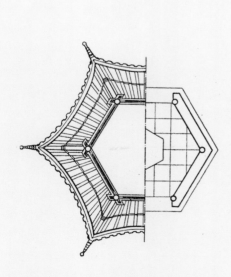

可亭平面及屋顶仰视图
Floor and ceiling plan of Ke Ting
(The Passable Pavilion)

可亭立面图
Elevation of Ke Ting
(The Passable Pavilion)

濠濮亭正立面图
Front elevation of Hao Pu Ting
(The Hao Pu Pavilion)

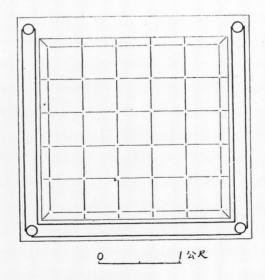

濠濮亭平面图
Floor plan of Hao Pu Ting
(The Hao Pu Pavilion)

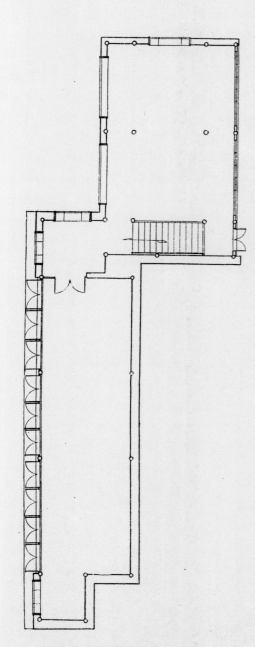

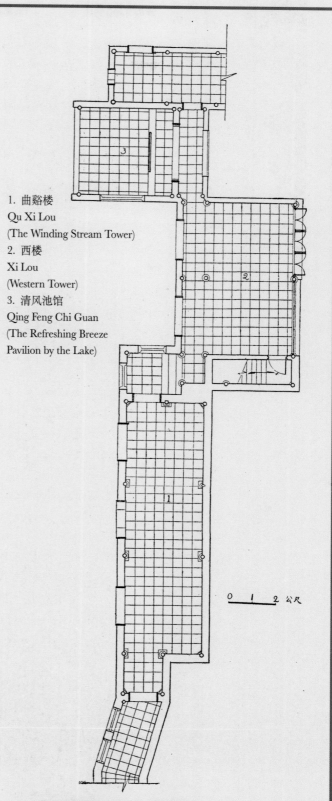

1. 曲豀楼
Qu Xi Lou
(The Winding Stream Tower)
2. 西楼
Xi Lou
(Western Tower)
3. 清风池馆
Qing Feng Chi Guan
(The Refreshing Breeze Pavilion by the Lake)

曲豀楼西楼二层平面图
Floor plan of second floor of Qu Xi Lou
(The Winding Stream Tower)
and Xi Lou (Western Tower)

曲豀楼西楼清风池馆底层平面图
Floor plan of ground floor of Qu Xi Lou (The Winding Stream Tower), Xi Lou (Western Tower) and Qing Feng Chi Guan (The Refreshing Breeze Pavilion by the Lake)

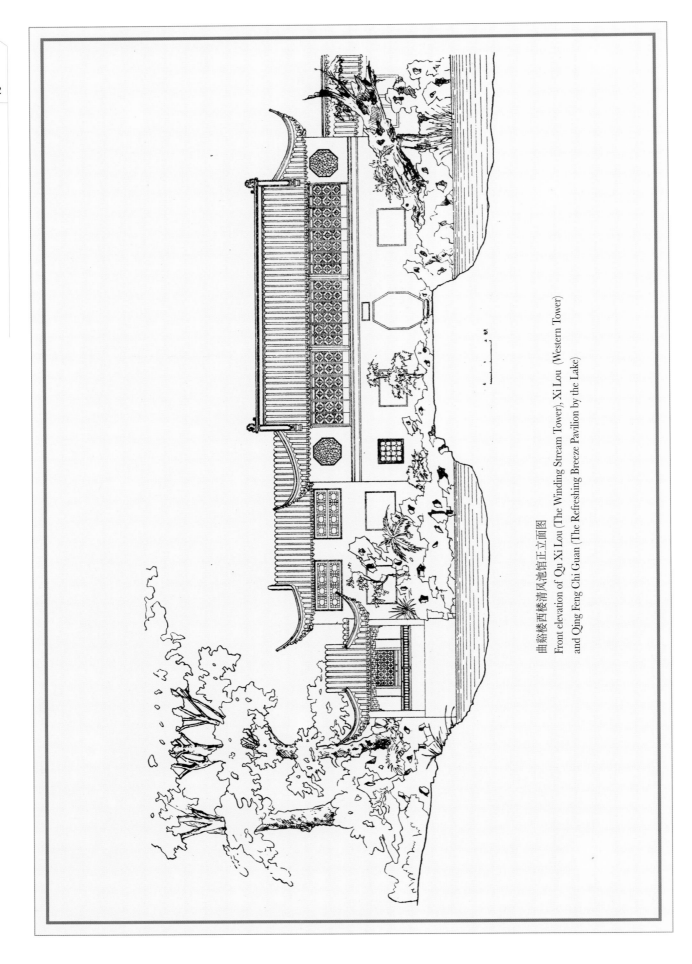

曲谿楼西楼清风池馆正画图
Front elevation of Qu Xi Lou (The Winding Stream Tower), Xi Lou (Western Tower) and Qing Feng Chi Guan (The Refreshing Breeze Pavilion by the Lake)

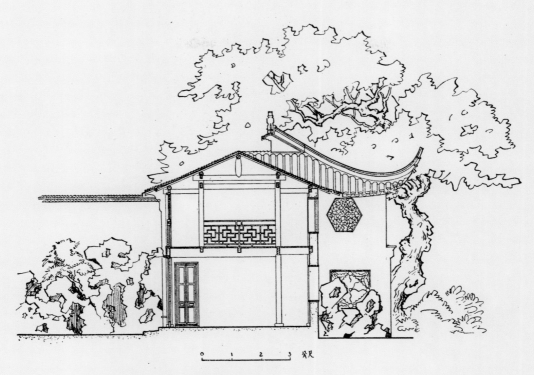

西楼横断面图
Cross section of Xi Lou (Western Tower)

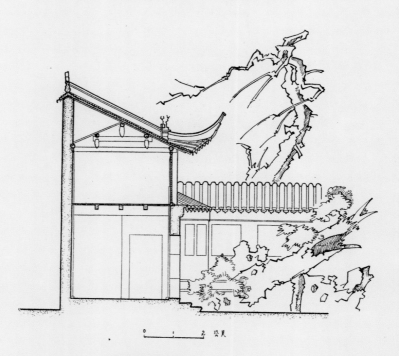

曲谿楼横断面图
Cross section of Qu Xi Lou
(The Winding Stream Tower)

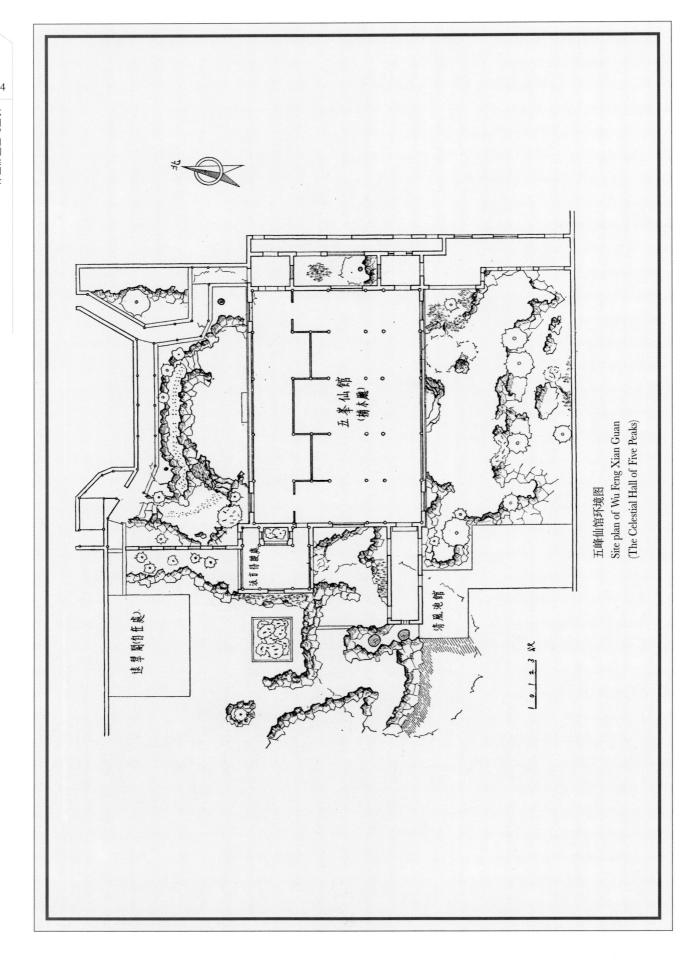

五峰仙馆环境图
Site plan of Wu Feng Xian Guan
(The Celestial Hall of Five Peaks)

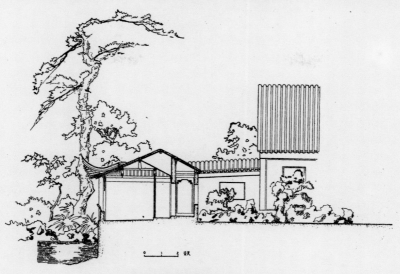

清风池馆横断面图
Cross section of Qing Feng Chi Guan
(The Refreshing Breeze Pavilion by the Lake)

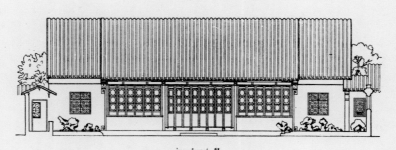

五峰仙馆背立面图
Back elevation of Wu Feng Xian Guan
(The Celestial Hall of Five Peaks)

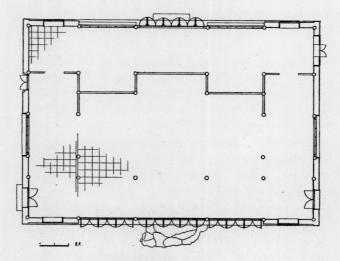

五峰仙馆平面图
Floor plan of Wu Feng Xian Guan
(The Celestial Hall of Five Peaks)

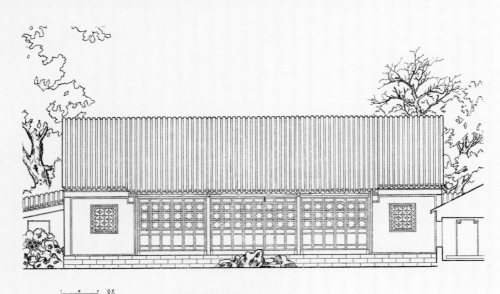

五峰仙馆正立面图
Front elevation of Wu Feng Xian Guan
(The Celestial Hall of Five Peaks)

五峰仙馆横断面图
Cross section of Wu Feng Xian Guan
(The Celestial Hall of Five Peaks)

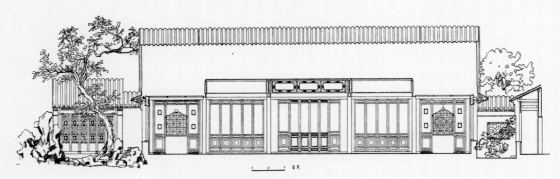

五峰仙馆纵断面图
Longitudinal section of Wu Feng Xian Guan
(The Celestial Hall of Five Peaks)

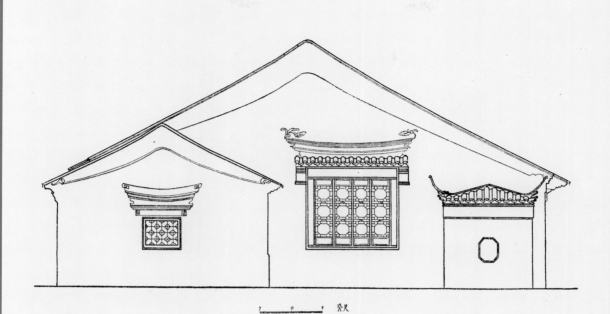

五峰仙馆侧立面图
Side elevation of Wu Feng Xian Guan
(The Celestial Hall of Five Peaks)

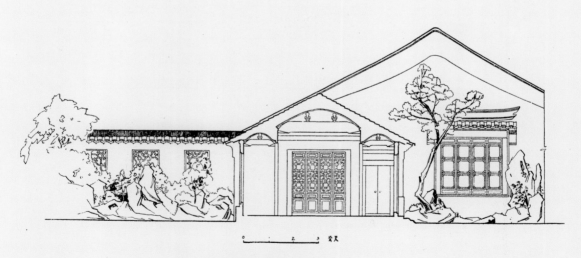

汲古得绠处横断面图
Cross section of Ji Gu De Geng Chu
(The Study of Enlightenment)

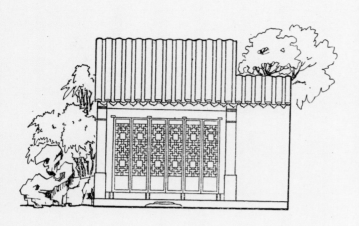

汲古得绠处正立面图

Front elevation of Ji Gu De Geng Chu
(The Study of Enlightenment)

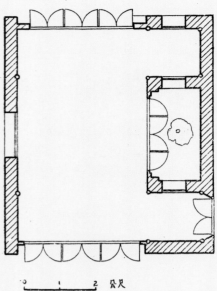

汲古得绠处平面图

Floor plan of Ji Gu De Geng Chu
(The Study of Enlightenment)

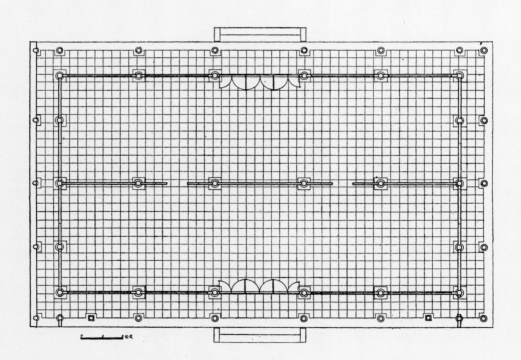

林泉耆硕之馆平面图

Floor plan of Lin Quan Qi Shuo Zhi Guan
(The Old Hermit Scholars' Hall)

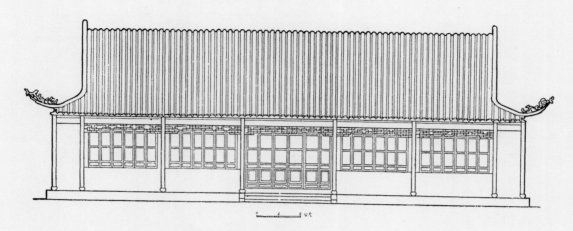

林泉耆硕之馆正立面图
Front elevation of Lin Quan Qi Shuo Zhi Guan
(The Old Hermit Scholars' Hall)

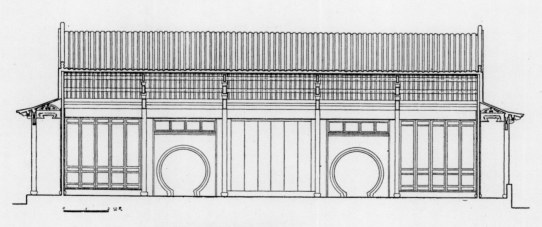

林泉耆硕之馆纵断面图
Longitudinal section of Lin Quan Qi Shuo Zhi Guan
(The Old Hermit Scholars' Hall)

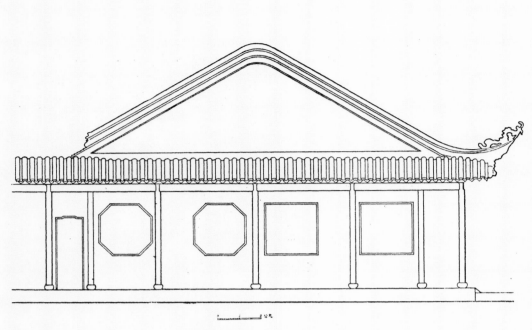

林泉耆硕之馆侧立面图
Side elevation of Lin Quan Qi Shuo Zhi Guan
(The Old Hermit Scholars' Hall)

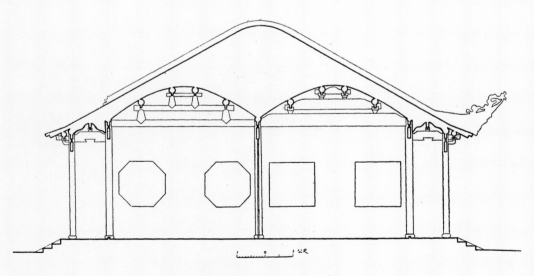

林泉耆硕之馆横断面图
Cross section of Lin Quan Qi Shuo Zhi Guan
(The Old Hermit Scholars' Hall)

亦不二亭横断面图
Cross section of Yi Bu Er Ting

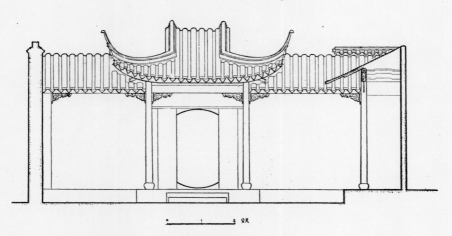

亦不二亭正立面图
Front elevation of Yi Bu Er Ting

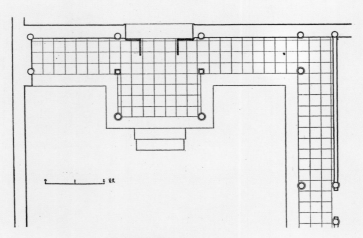

亦不二亭平面图
Floor plan of Yi Bu Er Ting

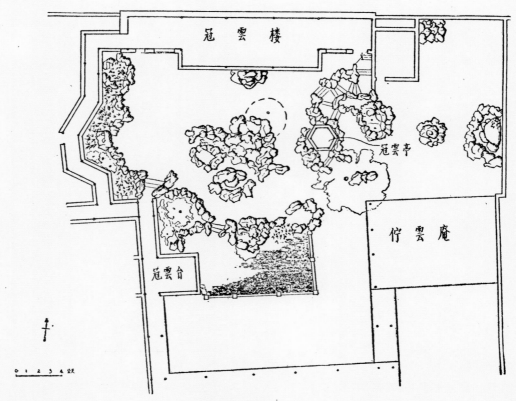

冠云楼环境图
Site plan of Guan Yun Lou
(The Cloud-Capped Tower)

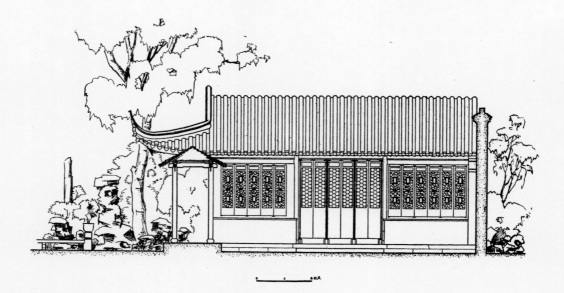

伫云庵正立面图
Front elevation of Zhu Yun An

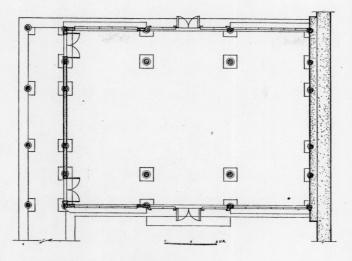

仁云庵平面图
Floor plan of Zhu Yun An

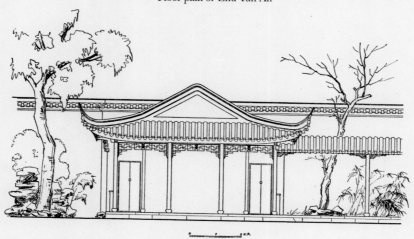

仁云庵侧立面图
Side elevation of Zhu Yun An

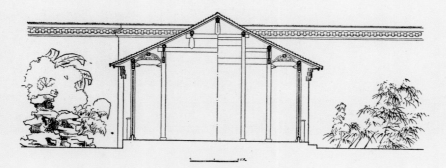

仁云庵横断面图
Cross section of Zhu Yun An

冠云亭正立面图
Front elevation of Guan Yun Ting
(The Cloud-Capped Pavilion)

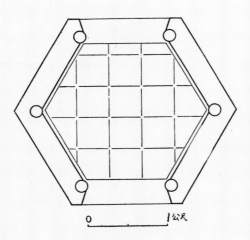

冠云亭平面图
Floor plan of Guan Yun Ting
(The Cloud-Capped Pavilion)

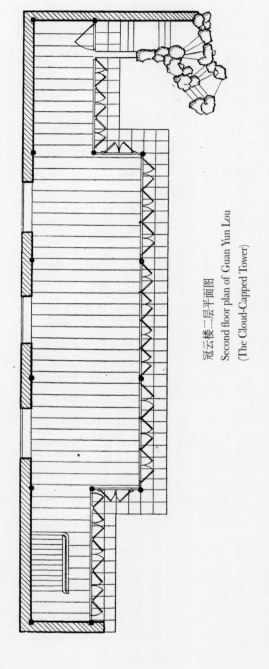

冠云楼二层平面图
Second floor plan of Guan Yun Lou
(The Cloud-Capped Tower)

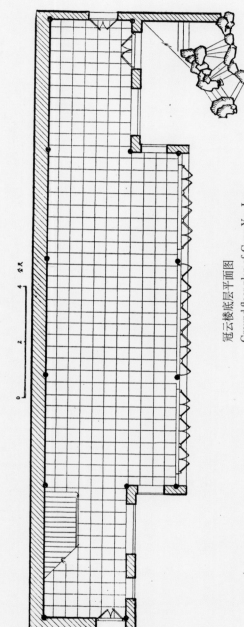

冠云楼底层平面图
Ground floor plan of Guan Yun Lou
(The Cloud-Capped Tower)

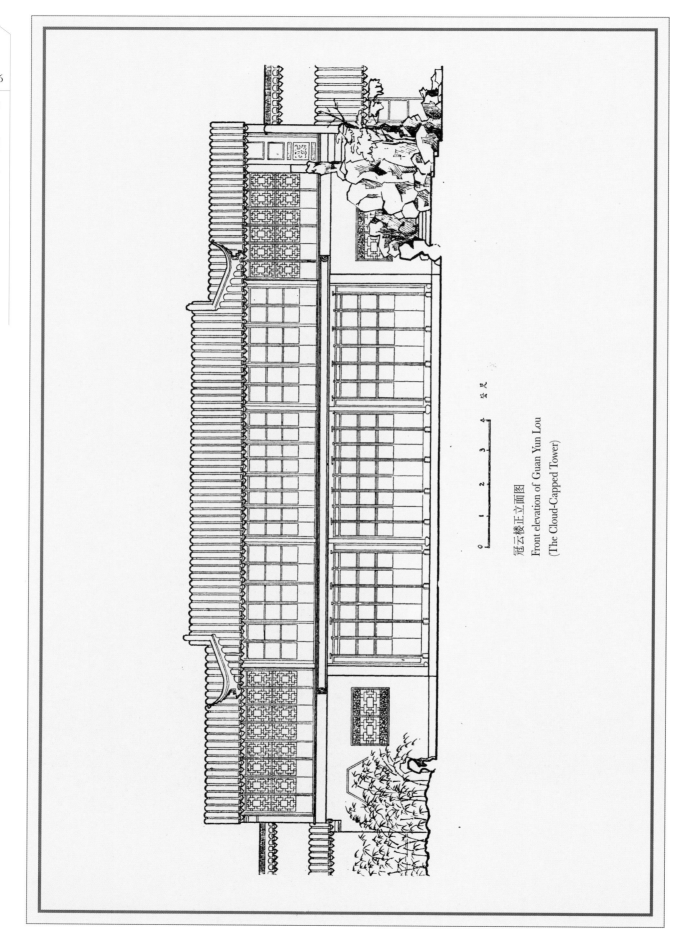

冠云楼正立面图
Front elevation of Guan Yun Lou
(The Cloud-Capped Tower)

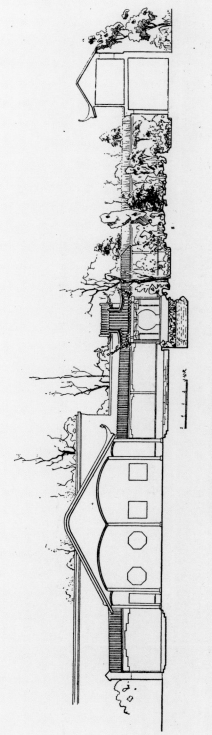

林泉耆硕之馆及冠云楼横断面图
Cross section of Lin Quan Qi Shuo Zhi Guan (The Old Hermit Scholars' Hall) and Guan Yun Lou (The Cloud-Capped Tower)

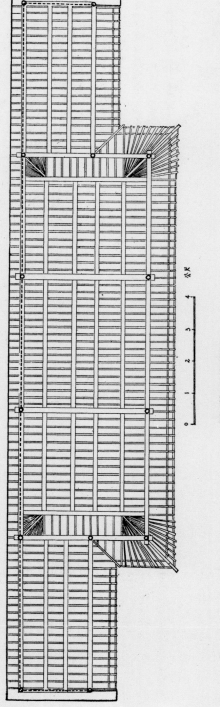

冠云楼屋顶仰视图
Ceiling plan of Guan Yun Lou (The Cloud-Capped Tower)

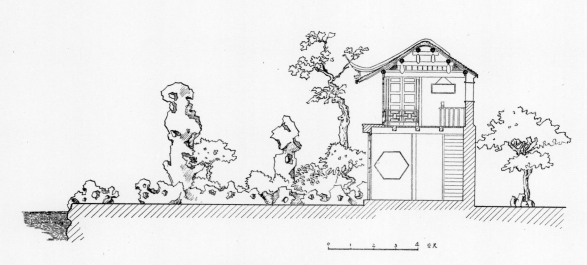

冠云楼横断面图
Cross section of Guan Yun Lou
(The Cloud-Capped Tower)

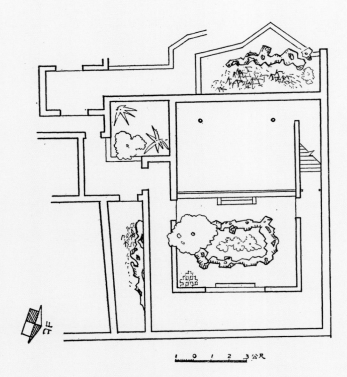

还我读书处环境图
Site plan of Huan Wo Du Shu Chu
(The Return-to-Read Study)

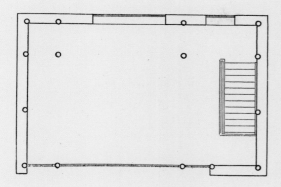

还我读书处二层平面图
Second floor plan of Huan Wo Du Shu Chu
(The Return-to-Read Study)

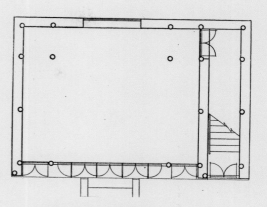

还我读书处底层平面图
Ground floor plan of Huan Wo Du Shu Chu
(The Return-to-Read Study)

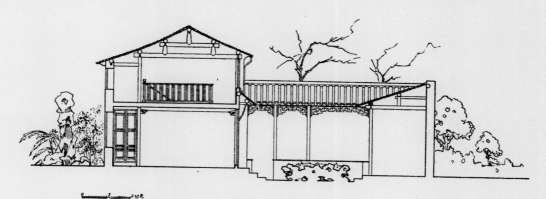

还我读书处横断面图
Cross section of Huan Wo Du Shu Chu
(The Return-to-Read Study)

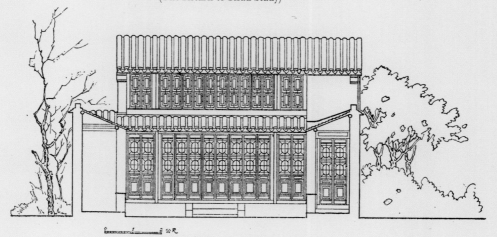

还我读书处正立面图
Front elevation of Huan Wo Du Shu Chu
(The Return-to-Read Study)

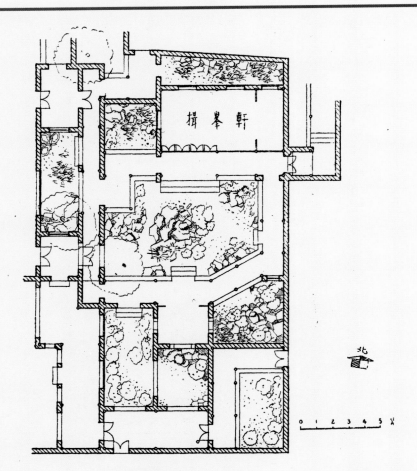

揖峰轩环境图
Site plan of Yi Feng Xuan
(The Worshipping Stone Pavilion)

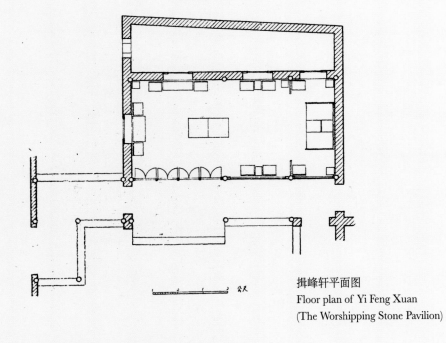

揖峰轩平面图
Floor plan of Yi Feng Xuan
(The Worshipping Stone Pavilion)

揖峰轩横断面图
Cross section of Yi Feng Xuan
(The Worshipping Stone Pavilion)

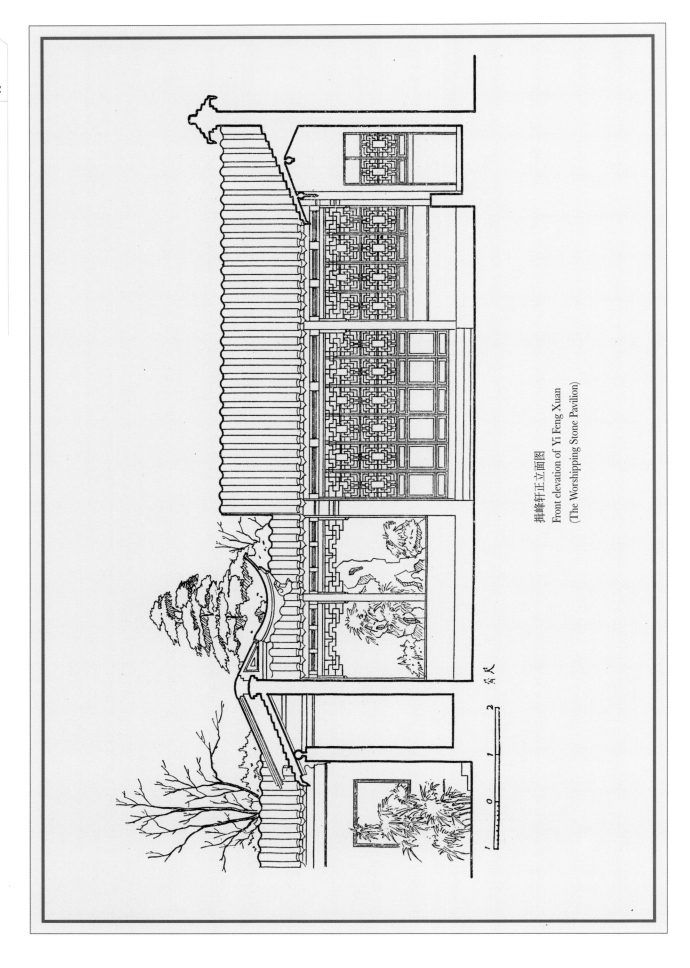

揖峰轩正面图
Front elevation of Yi Feng Xuan
(The Worshipping Stone Pavilion)

冠云台正立面图
Front elevation of Guan Yun Tai

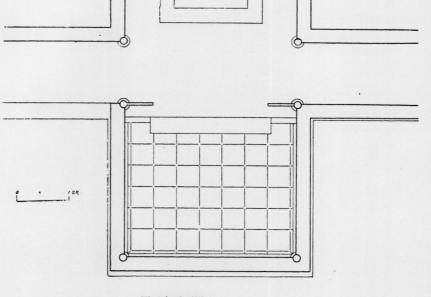

冠云台平面图
Floor plan of Guan Yun Tai

佳晴喜雨快雪之亭及冠云台横断面图
Cross section of Jia Qing Xi Yu Kuai Xue Zhi Ting
(The Good-For-Farming Pavilion) and Guan Yun Tai

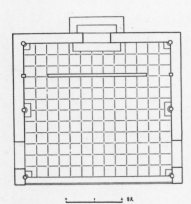

佳晴喜雨快雪之亭平面图
Floor plan of Jia Qing Xi Yu Kuai Xue Zhi Ting
(The Good-For-Farming Pavilion)

佳晴喜雨快雪之亭正立面图
Front elevation of Jia Qing Xi Yu Kuai Xue Zhi Ting
(The Good-For-Farming Pavilion)

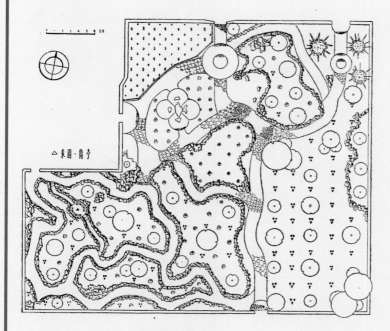

东园一角亭环境图
Site plan of a pavilion at the corner of eastern portion

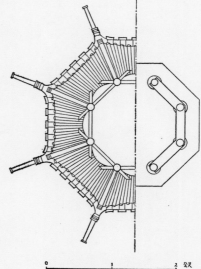

东园一角亭平面及屋顶仰视图
Floor and ceiling plan of a pavilion at the corner of eastern portion

东园一角亭立面图
Elevation of a pavilion at the corner of eastern portion

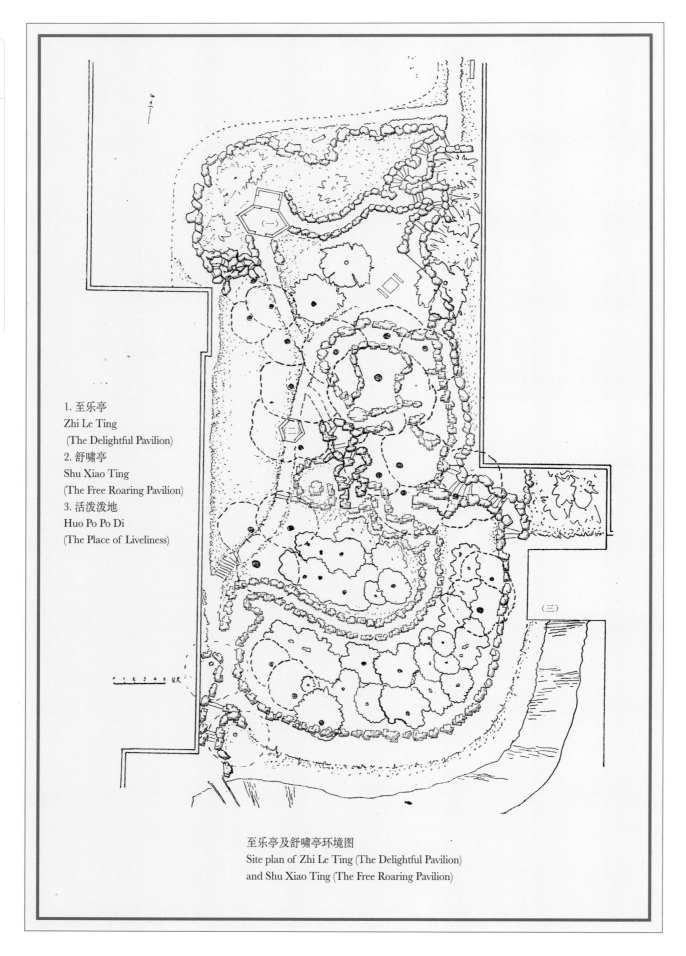

至乐亭及舒啸亭环境图
Site plan of Zhi Le Ting (The Delightful Pavilion) and Shu Xiao Ting (The Free Roaring Pavilion)

1. 至乐亭
   Zhi Le Ting
   (The Delightful Pavilion)
2. 舒啸亭
   Shu Xiao Ting
   (The Free Roaring Pavilion)
3. 活泼泼地
   Huo Po Po Di
   (The Place of Liveliness)

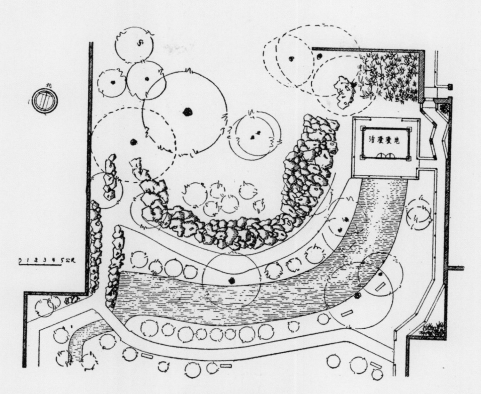

活泼泼地环境图
Site plan of Huo Po Po Di
(The Place of Liveliness)

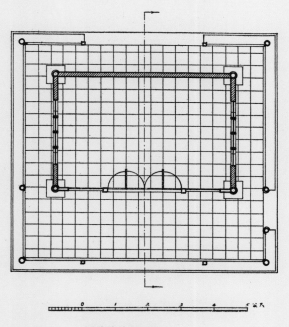

活泼泼地平面图
Floor plan of Huo Po Po Di
(The Place of Liveliness)

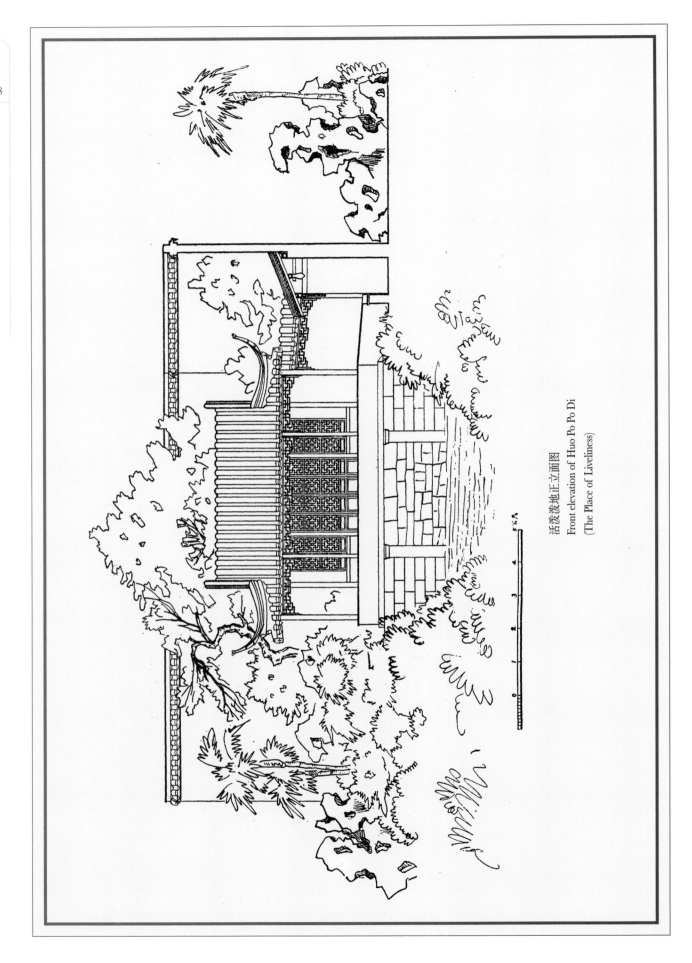

活泼泼地正立面图
Front elevation of Huo Po Po Di
(The Place of Liveliness)

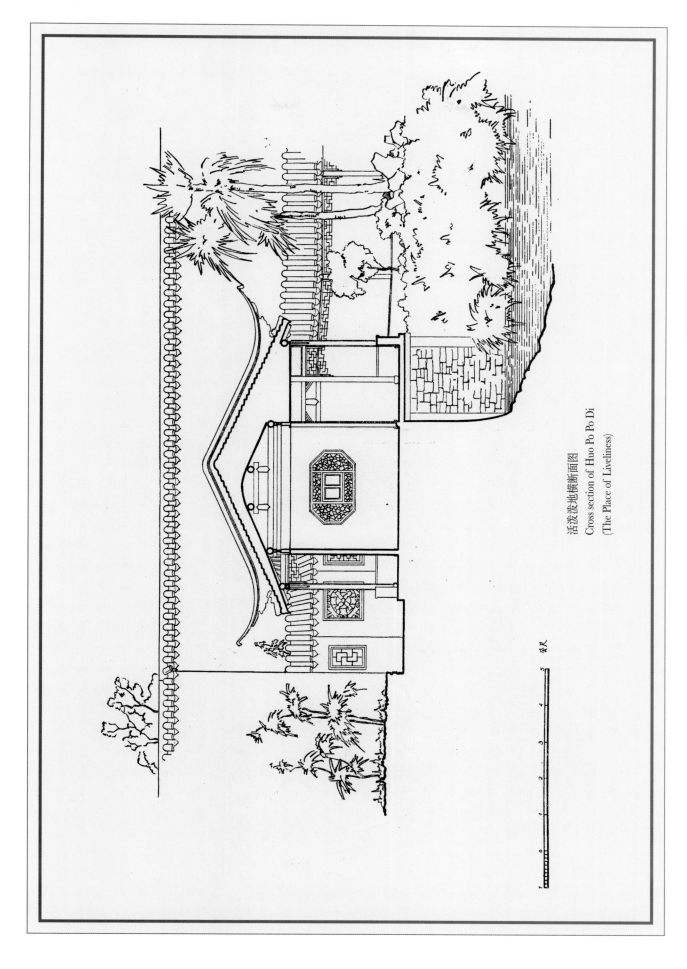

活泼泼地横断面图
Cross section of Huo Po Po Di
(The Place of Liveliness)

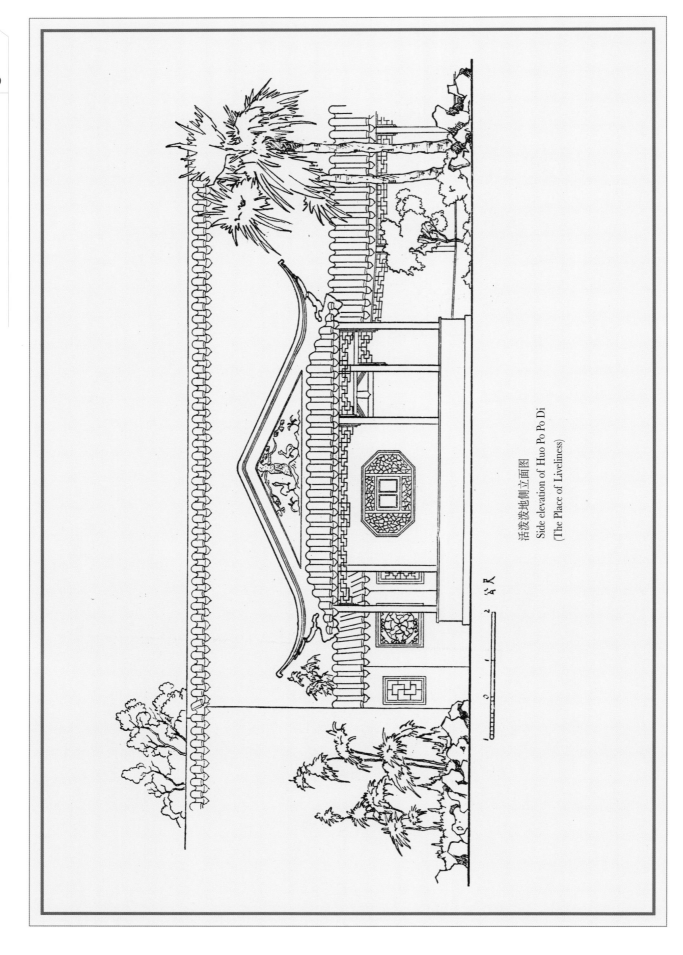

活泼泼地侧立面图
Side elevation of Huo Po Po Di
(The Place of Liveliness)

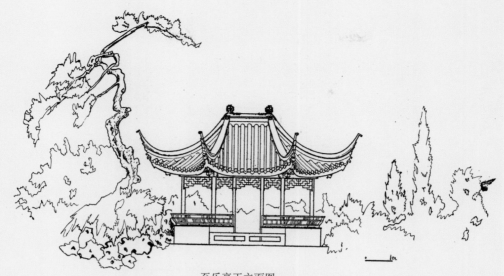

至乐亭正立面图
Front elevation of Zhi Le Ting
(The Delightful Pavilion)

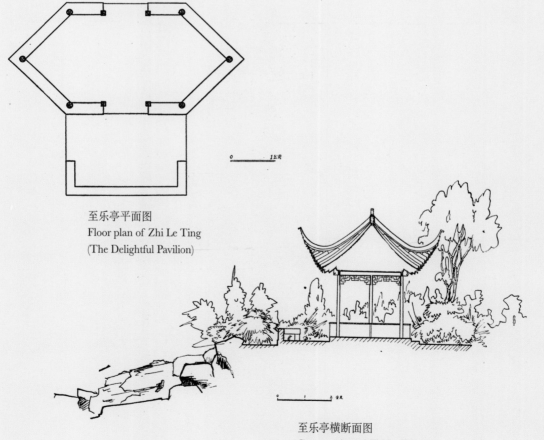

至乐亭平面图
Floor plan of Zhi Le Ting
(The Delightful Pavilion)

至乐亭横断面图
Cross section of Zhi Le Ting
(The Delightful Pavilion)

舒啸亭平面图
Floor plan of Shu Xiao Ting
(The Free Roaring Pavilion)

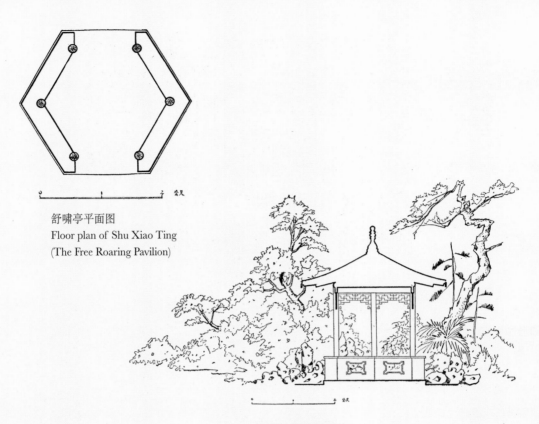

舒啸亭横断面图
Cross section of Shu Xiao Ting
(The Free Roaring Pavilion)

舒啸亭立面图
Elevation of Shu Xiao Ting
(The Free Roaring Pavilion)

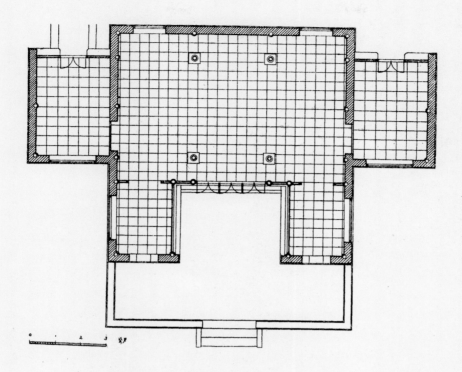

陈列馆平面图
Floor plan of Exhibition Hall

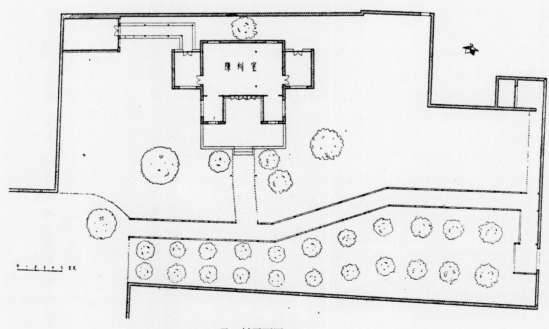

又一村平面图
Site plan of You Yi Cun

测绘图录编后附记

　　一九五六年五月，我率领同济大学建筑系建筑学专业二年级同学到苏州无锡进行教学实习，主要对象是参观新旧及古典建筑，尤其对于苏州的园林我们又作为重点，在实习中进行了测绘，本集中的测绘图即该级同学的成绩的一部分。参加指导实习的除我兼总领队外，尚有我系教师王志英，朱保良，陈光贤三君，都尽了很多的力。封面图案是张孚珩同学的匠心，特此一并附记。

　　参加同学名单：

沈在安	陈世杰	殷焕圻	张之俊	马光蓓	杨维良	沈鼎铭	朱有琳
曹庆涵	陈荣萱	焦璞文	张纪延	黄宏骥	蒋守谦	叶佐豪	陈崎
程弋日	刘纪芬	张景新	郑定国	王慧英	郁学儒	莫琴	陈明
郑国英	赵群娇	龙永龄	郁操政	高正秋	沈瑞莲	沈尧	张华
李鑫林	赵凤珍	王周云	唐班如	林言官	叶丹霞	陶祥兴	沈惠骥
胡圣儒	顾琪美	金福珍	钟临通	谭华爵	施家谱	殷鉴明	冯渭文
朱培兰	郑丽菊	苏邦俊	张聿洁	安怀起	艾亨音	戴天军	朱谋隆
曹书安							

一九五六年十月　陈从周

Notes

　　In May, 1956, I led the second year students with a major of Architects in Architecture Department of Tongji University to take a learning practice in the cities of Suzhou and Wuxi. Our main activities were visiting the traditional buildings in these two cities. In particular, we did an architectural survey on the gardens in Suzhou. The survey drawings collected in this book are a part of the result of this learning practice. Three other teachers, Wang Zhiying, Zhu Baoliang, Chen Guangxian have spent great effort in this practice. The drawing at the book face was made specially by the student Zhang Fuheng.

Noted by Chen Congzhou in October, 1956.

后 记

中英文的新版《苏州园林》，在大家的鼎力相助下，今日终于问世了。她距离原版"苏州园林"的诞生，岁月已足足逝过了半个多世纪。"写得宏图欣得意，繁花开遍阖闾城"是父亲所作的"吴门杂咏"中的两句诗，表达了父亲对苏州寄予的厚望。我们希望新书的出版可弘扬中华的传统文化并告慰九泉之下的父亲。

父亲陈从周先生是中国古典园林建筑学家，生前是同济大学建筑系教授，著有《说园》等许多园林著作，其中不少被译成英日德法意俄等文字，一版再版。在这些书中，父亲最钟爱的还是《苏州园林》，把其称作是自己的第一本书。这本书不仅是父亲的成名作，更是他的所有著作中最不同寻常的一本书。她奠定了父亲的园林建筑人生以及他在中国古典园林艺术界的地位，同时她也曾经给父亲带来苦难，使他饱受折磨。

这本书当时只不过是同济大学建筑系的一本教材，出版后却立即引起了社会上的注意。建筑史学界的老前辈高度评价它，同济师生们爱读它。据说当年叶圣陶先生看到这本书后，便立即想见父亲；在他想象中，陈从周应该是一个老者，不然怎么能写出这么一本书；认识父亲以后，才发现父亲当年只不过三十多岁，从此他俩便成了几十年的忘年交。

尽管这本书命运多舛，出版不久就被打入冷宫，但它让中国乃至世界进一步认识了苏州园林。凭借这本书，20世纪70年代末，纽约大都会博物馆找到了父亲，于是便有了以网师园殿春簃为蓝本的明轩，苏州园林从此走向了世界，也开始了父亲和贝聿铭先生近二十年的友谊。贝先生说："陈从周先生，中国园林艺术之一代宗师，仁人君子，吾之挚友。吾与从周初识于上世纪七十年代，恨相知晚也。每每聆听从周说园林、谈建筑、议评弹、论昆曲，甚为投机，畅须教益。得此知己，吾欣慰不已。"[①]

在父亲眼中，中国园林不是一门孤立的建筑艺术；它不仅有政治、经济的历史背景，还和绘画、诗词、戏曲息息相关。贝先生为之赞叹说："陈从周对中国园林之理解肌擘理分，博大精深，非凡人所能及。"

父亲曾师从张大千绘画，起先习工笔，后来攻写意。20世纪40年代开过个人画展，出过画集；其中一幅"一丝柳一寸柔情"，曾经一时蜚声沪上。画面上是一对羽毛初丰的小鸟,站在细细的柳枝上,唧唧我我地亲热着。看父亲作画是一种美的享受,研墨、摊纸、润笔,稍停片刻,手臂在空中飞舞,笔尖在宣纸上疾行,转而狂揿着墨汁,继而回到纸上跳跃。弯腰的兰、挺拔的竹、硕大的蕉叶,一一浮现在纸上。画上熟透了的葡萄快破皮了,他就赶忙把烟头移近,将漫开的水迹烘干。明代画家文徵明曾为拙政园作画许多,父亲也是以山水画的视角赏园品园。他认为"不知中国画理画论,难以言中国园林"②。父亲造园修园,用的也是绘画的手法。他强调"园以景胜,景以园异",构园"有法而无式"③。他不拘泥于图纸,因地制宜;在现场触景生情,"诗中有园,画中有园"④,天马行空的构思常常会心想手动地往纸上涂炭,往地上洒灰线。这种设计虽然奇特,但父亲做来却是有章有序。

父亲爱好摄影,照片拍得很有意境,他的绘画技巧把摄影技术发挥得淋漓尽致。我学拍照的时候,父亲就教我表现景物的细节要用小光圈、长曝光。为了在自然光下拍出园林建筑的神韵,父亲常常支起三脚架,按下快门,然后点上烟,坐在一边静静地等待曝光结束。当年父亲用的是一架德国双镜头120相机,是我四姨夫宋伯钦离开大陆前留下的。《苏州园林》书中一百九十六张的照片都是用这架相机拍的,叶圣陶先生称这些照片"全都是艺术的精品:这可以说是建筑界和摄影界的一个创举"⑤。

父亲当年就读之江大学文学系,师从夏承焘先生学词。父亲以文人的情怀为园林立意,别出新裁地撷取宋词为书中的照片题款。真是煞费苦心。撷句的择用出处广泛,为了让词句更能传神,甚至有些照片的上下句撷自不同的宋词,个别地方父亲还改动了个别字,从而起到了画龙点睛的作用。这些经过父亲再创作的文字与画面完美结合,可谓出神入化,融合得天衣无缝。

父亲喜欢听昆曲,高兴时还会唱上几句,并能吹箫。夜深人静时,悠扬的箫音和唱词常伴随儿时的我和弟妹进入甜蜜的梦乡。在父亲眼里,园林和昆曲同根生,在他园林著作的字里行间也充满着戏曲元素。曲要静听,园宜静观,观之才有得。园景和戏曲是不可分割的;中国园林中的一山一木,一榭一亭,犹如舞台上一举一动,一词一曲;而园林的韵律,曲折高下,又同昆曲无二致。所以父亲提倡要"以园解曲,以曲悟园"⑥。

就是这样,父亲的人文艺术和文学功底,在建筑领域里得到极大的发挥,他以画家的敏感,学者的严谨,考古工作者的刻苦精神,用现代建筑园林的眼光,中国古典文学的风格写成了这本《苏州园林》。宋代李格非写了本《洛阳名园记》,父亲当年凭借对苏州园林的热情也想学李格非,写一本《苏州名园记》,结果就写成了《苏州园林》。这是一部新形式的建设史学专著,书中完整的园林建筑测绘图,是当年父亲带领同济大学建筑系师生们手绘的。

《苏州园林》自1956年由同济大学教材科初版后,虽名扬四海,却发行量不多,只是1983年在东京以日文出版过一次。这次重新出版,我们决心全书英译;希望世人能籍此更好地欣赏与理解灿烂的苏州园林文化。

新版本的完稿凝聚了两代人的艰苦努力。父亲的文章就像他的画,下笔简练,含意丰富。通篇文章娓娓道来,温雅婉约,间而或几句或大段的抒情,诗一般的文字,叹为观止。文中还有大量的古建筑专业词汇,并涉及许多历史人物与文献。所以要把这本涉及历史、建筑、文学、绘画诸领域的著作传神地译成英文,实在不容易。陈威虽然过去已经译过《苏州旧住宅》,这次翻译书中"苏州园林的初步分析"一文仍颇感困难。译文经过北京语言大学朱文俊教授反复修改润色后,美妙而准确地表达了原文的含义,起到了美轮美奂的效果。朱文俊教授曾在联合国教科文组织长期担任英译审校工作,去年再版的英译稿《说园》也是请他校审的。

更为难能可贵的是书中照片的宋词英译。首先,为了寻找原书中这些宋词撷句的出处,张熹先生用现代网络搜索技术,搜索了近三万首宋词才寻获这些宋词出处,为文章的翻译奠定了基础,其工作之艰辛,足可以写成一部书。接下来的英文翻译就更具有挑战性了。朱文俊教授几易其稿,力图在意、音、形上都达到唯美的境界。我们希望读者们在阅读中会有所感受。新版《苏州园林》中每幅照片的宋词撷句,仍由蒋启霆先生缮写,他曾为父亲的《说园》用小楷抄录了全文。

另外值得一提的是,本书的完成也称得上是一个群策群力的结果,得到不少专家和热心人的帮助。仅用"感谢"两个字是难以表达我们由衷感激的。美国加州伯克莱大学李林德女士翻译过昆曲《牡丹亭》，她多次来信指点我们从何处着手。张熹不仅查找出宋词撷句的出处，还对书中的一些典故作了考证。为了规范景点译名，苏州马汉鹏和曾澄夫妇几乎跑遍了苏州园林，找来了所有能找到的书中的景点译名，还实地拍了许多照片，提供了有用的资料。苏州园林设计院有限公司贺凤春女士特地寄来苏州园林局撰写的《苏州古典园林》一书。徐正平费心地修改这篇后序。

父亲去世已将近十年，我们为他做了几件事，包括筹建南北湖陈从周艺术馆、编辑《陈从周画集》、出版《苏州旧住宅》和《园综》等等。从《苏州园林》初版到今天的新版，半个多世纪过去了，中国发生了巨大变化。寻找历史、留住文化的想法，激发了我们对《苏州园林》一书进行整理翻译。历时两年的艰辛工作把我们重新拉到了父亲的身边，听园林、闻箫笛，又一次得到美的熏陶，真是受益匪浅，我们深深地怀念和感谢父亲！

陈胜吾

二〇〇八年八月

注：
① 《陈从周传·序》，上海文化出版社 2009 年版。
② 《园林与山水画》，陈从周，《陈从周散文》，同济大学出版社 1999 年版。
③ 《中国的园林艺术和美学》，陈从周，《陈从周散文》，同济大学出版社 1999 年版。
④ 《春苔集》序，陈植，同济大学出版社 1985 年版。
⑤ 《苏州园林》序，叶圣陶，1979 年 2 月 6 日，《叶圣陶散文乙集》，三联书店 1984 年版。
⑥ 《以园解曲，以曲悟园》，陈从周，《陈从周散文》，同济大学出版社 1999 年版。

Afterwords

The new bilingual version (Chinese and English) of *Suzhou Gardens*, with the full back-up of all my friends, has come to light. It is almost over half a century since the publication of the original version of the book.

"I paint the wonderland with pride and joy, To let flowers flourish all over Suzhou"

These are two poetic sentences from my father's *A Collection of Poems on Suzhou*, expressing his great expectation for the future of Suzhou. It is our hope that the new book will uphold the cultural heritage of China and bring consolation to our dear deceased father.

Mr. Chen Congzhou, my father, was a master architect of traditional gardens of China. He used to be a professor of the Architecture Department of Tongji University. He wrote numerous books and articles on gardens, including *On Gardens*. Of all the writings, some have been published in English, Japanese, German, French, Italian, Russian and other foreign languages, and witnessed several printings or editions. But my father's favorite book is *Shuzhou Gardens*, his first book, as he called it. This book is really a very unusual one. On one hand, it makes my father's name nationwide, establishing his position in the history of garden construction and the field of art of traditional Chinese gardens. On the other hand, it brought him many unexpected troubles and inflicted great suffering on him.

In fact, the original version of this book was only a course book used in the Architecture Department of Tongji University. Since its publication, it attracted widespread attention in the society. Senior doyens of architecture history had a high appraisal of the book. The teachers and students loved reading it. It is said that when Mr. Ye Shengtao (叶圣陶), the renowned scholar, had read this book, he wanted to see my father immediately. In Ye's mind, the person who is able to write such a wonderful book must be a senior scholar. To Ye's astonishment, he found when they finally met each other that my father was only a youth of over thirty years of age at the time of writing that book. Ever since, they became friends for dozens of years in spite of their age gap.

This book has deepened the awareness of China and the world of the gardens of Suzhou, though it was banished soon after its publication and then had a precarious fate. Through this book, the Metropolitan Museum of New York called on my father at the end of 70s of the last century to design the *Ming Xuan* (Ming

Room, 明轩, or the Astor Court) in the Museum, based on *Dian Cun Yi* (Late Spring Studio, 殿春簃) in *Wang Shi Yuan* (The Master-of-nets Garden, 网师园). From then on Suzhou gardens made their name in the world. At the same time, my father struck up friendship of nearly twenty years with Mr. I. M. Pei (贝聿铭), the world-famous architect. Mr. I. M. Pei said: "Mr. Chen Congzhou is a great master of his time in garden art. He is a benevolent gentleman and a good friend of mine. I got to know him in the 70s of the last century. I regret not having met him earlier." "Having such an intimate friend, I am immensely gratified."

In the mind of my father, Chinese garden construction is not an isolated architectural art. It is not only involved with the historical background of politics and economy, but also related with painting, poetry and drama. Mr. I. M. Pei speaks highly of my father: " Mr. Chen's understanding of Chinese garden art is thorough, profound and far above the pale of the average person."

My father studied painting under Zhang Daqian (Chang Dai-chien, 张大千). He learned *gongbi* (a traditional Chinese painting method featured by fine brush work and detailed description) at first. And then he turned to the painting style of freehand brushwork. In the 40s of the 20th century, he held his personal painting exhibition and had an album of paintings published. One of the paintings entitled "A Thread of Willow, A Length of Tenderness" once enjoyed high reputation in Shanghai. The painting features a pair of fledgling little birds perching on a thin willow branch and chirping sweetly to each other. To watch my father in the act of painting is indeed an enjoyment of beauty. He started by rubbing an ink stick against an inkstone, spread a large piece of Xuan paper and dipped his brush into the ink. After a little pause, he began to wield his arm in the air and let the tip of his brush sliding swiftly on the paper. Then he dipped his brush again into the ink and brought it back to continue its dancing on the paper. The bending orchid, the upright bamboo, the giant palm leaves all magically emerged on the paper. Finding that the ripe grapes on the painting seemed to burst, he hastened to bring a burning cigarette near them to dry up some overspreading ink water. Mr. Wen Zhenming, a famous painter of the Ming Dynasty, did numerous paintings for *Zhuo Zheng Yuan* (The Humble Administrator's Garden, 拙政园). In the same way, my father appraised and appreciated gardens from the angle of Chinese landscape painting. He held that "without knowledge of the mechanism of Chinese painting, no comments on Chinese gardens are justified." In the construction and reparation of gardens, he also resorted to the methodology of painting. He emphasized that "A garden excels in its landscape and landscape prides in being distinctive." and "there is no ready-made formula to follow in building a garden though there are some general guide lines to go by." In fact, he did not stick rigidly to the blueprint. And instead he often adapted his measures in light of local situations or made impulsive improvements at the sight of actual scenes. He advocated "garden is embedded in both poetry and painting". He scrawled and splashed the sketch of a scene on paper and even sprinkled lime lines on the ground as dictated by flights of imagination. This way of designing seemed unusual but accorded with his own principles and procedures.

My father was keen on photography. The pictures he shot were full of import. The modern technique of photography gave full play to his profound understanding of Chinese painting. When I learnt picture-taking, my father taught me how to use small aperture and long exposure to display the details of things I shot. In order to turn out the charm of garden buildings, my father often put up the tripod and clicked the shutter. And then he sat down, lighting up a cigarette and quietly waiting for the end of exposure. In those years, my father used a German-made 120 film twin-lens reflex camera given him by my uncle Mr. Song Boqin when the latter left Chinese mainland. All the 196 pictures carried in the book *Suzhou Gardens* were produced by this camera. Mr. Ye Shengtao regarded these pictures as "all fine pieces of art and an pioneering undertaking in the fields of architecture and photography".

My father is diligent and well-grounded in literature, as he studied humanities at Zijiang University under Mr. Xia Chengxi. Inspired by the sentiments of a man of letters, he inscribed his pictures with words ingeniously and painstakingly selected from Song ci (poems of the Song Dynasty). His choice of poetic words came from wide-ranging sources and sometimes two consecutive sentences were quoted from two different Song poems in order to enhance their expressiveness. Occasionally he replaced one or two particular words to bring the poem and the picture to life. These recreated words matched the pictures of the book to a charm and attained the acme of perfection.

My father was fond of *Kunqu opera*. When he was in a happy mood, he could even sing a few stanzas. He was a skillful player of Xiao (a vertical bamboo flute). In the day of yore, when the night was deep and quiet, the melodious music of the flute and the ballad often took my younger brother and sister and I into the sweet dreamland. In the eye of my father, gardens and *Kunqu opera* grew up from the same root. There were elements of drama running between the lines of his works on gardens. Music should be enjoyed when listened with bated breath; gardens should be appreciated when observed quietly and calmly for their best aesthetic effect. Melody and landscape are inseparable. It seemed to him that the hills, trees, pavilions and pagodas in a garden are just like the poses, movements, words and tunes of the stage art. In the same way, the undulating rhythm of gardens synchronizes with *Kunqu opera*. With this understanding in mind he upheld all along the idea of "decoding music with the aid of gardens and reading the gardens by means of music."

In this way, my father's solid grounding and talents in art and literature were brought into full play in modern garden architecture. It was with the sensibility of a painter, the rigor of a scholar, the diligence of an archeologist, the version of modern garden architecture and the style of Chinese classic literature that he brought the book *Suzhou Gardens* into the world. Mr. Li Gefei of the Song Dynasty wrote a book entitled *An Account of Loyang Gardens*. My father, on the strength of his love for Suzhou gardens, wanted to learn from Mr. Li and to write a book *An Account of Suzhou gardens*. But in the end his efforts landed the present book *Suzhou Gardens*. This is an update monograph on the history of architecture. The complete set of survey maps of the gardens carried in the book are hand-made by the teachers and the students of the Architecture Department of Tongji University under the guidance of my father.

Since its publication by the Course-book Press of Tongji University, the book *Suzhou Gardens* became well-known far and wide, but its circulation was limited. Later in 1983, the Japanese version of the book was published in Tokyo. This time we decide to have the whole book translated into English, in the hope that the new version might help people nowadays to better understand and appreciate the brilliant garden culture of Suzhou.

The new version is the brainchild of the pains-taking efforts of two generations. My father's essays, like his paintings, are pithy, lucid and rich in meaning. The essays are characterized by an urbane and graceful narrative style and occasional passages, big or small, of lyric expressions. These essays also feature large loads of technical words of ancient architecture and of allusions to historical figures and documents. Due to the wide coverage of fields such as history, architecture, literature, painting and so on, to have the book translated into English adequately and vividly is by no means an easy matter. Mr. Chen Wei did the translation of the book *The Old Houses of Suzhou* a couple of years before, but to translate the essay *A Preliminary Analysis of Suzhou Gardens* in the present book is still a challenge. His English version has been revised and polished by Prof. Zhu Wenjun of Beijing Language and Culture University. The finalized English version comes up faithfully with the original meanings, making the whole book a wonderful reading. Prof. Zhu worked for UNESCO for several years as a reviser. Last year he helped revise the English version of the second edition of my father's book *On Gardens*.

The English version of Song ci attached to the pictures is worthy of particular mention. First of all, tremendous efforts were made to find out the sources of all those quoted Song ci sentences. Mr. Zhang Xi availed himself of modern network technique for this challenging job. In the end he succeeded in ferreting out all the sources from among nearly 30,000 Song ci, thus paving the way for the translation. Even more challenging part is the translation. Prof. Zhu revised his English version again and again in an attempt to perfect it as much as possible in meaning, sound and form. It is our hope that the readers may appreciate what has been done on our part and enjoy reading it. All the quoted Song ci sentences in the current version were transcribed by late Mr. Jiang Qiting, who had done the transcription for the whole text of the book *On Gardens* in its year 1984 edition.

It is also worthwhile to mention that the publication of this book is the fruit of concerted efforts and generous assistance of some experts and helpful friends. To them we express our heartfelt and sincere appreciation. Prof. Li Deling of UC Berkely, who translated Kunqu opera *Peony Pavilion*, wrote to me several times about how to proceed with the work. Mr. Zhang Xi not only sought out all the sources of the quoted Song ci sentences, but also made some textual research on a number of allusions in the book. Mr. Ma Hanpeng and his wife Mrs. Zeng Cheng visited almost all the gardens in Suzhou and collected all English names for scenic spots. They also took many photos on the spot to provide data and clues for later reference. Miss He Fengcung of the Limited Company of Suzhou Institute of Garden Designing sent by mail a book *Classic Gardens of Suzhou* written by the Garden Bureau of Suzhou. Mr. Xu Zhengping took pains revising the *Afterwords* part.

It has been almost a decade since my father passed away. In honor of his memory, we have done several things such as the preparation of the Nanbei Lake Art Gallery of Chen Congzhou, editing *Painting Album of Chen Congzhou*, having *The Old Houses of Suzhou* and *An Overview of Gardens* published, and so on. Over half a century has elapsed from the original to the present new version of *Suzhou Gardens*. During this period of time China has undergone profound changes. The thought of recalling the history and reviving the culture has inspired us to edit and translate the book *Suzhou Gardens*. Two years of persistent and painstaking efforts have drawn us to our father's side, to listen to his melodious flute and his enchanting stories about gardens. We feel so happy and so graceful to be nurtured again by his delicate sense of beauty. We are missing him so much. We will cherish the memory of our dear father forever.

Chen Shengwu
Oct. 2008

图书在版编目（CIP）数据

苏州园林：汉英对照/陈从周著. —上海：上海
人民出版社,2018
ISBN 978-7-208-15282-3

Ⅰ.①苏… Ⅱ.①陈… Ⅲ.①古典园林-园林艺术-
苏州-汉、英 Ⅳ.①TU986.625.33

中国版本图书馆 CIP 数据核字（2018）第 138942 号

策　　划	陈胜吾　陈　馨
翻　　译	朱文俊　陈　威
责任编辑	苏贻鸣　张晓玲
美术编辑	溪　远

苏州园林

（汉英对照）

陈从周　著

出　　版	上海人民出版社
	（200001　上海福建中路 193 号）
发　　行	上海人民出版社发行中心
印　　刷	上海盛通时代印刷有限公司
开　　本	890×1240　1/16
印　　张	24
插　　页	6
字　　数	361,000
版　　次	2018 年 8 月第 1 版
印　　次	2018 年 8 月第 1 次印刷
ISBN	978-7-208-15282-3/K・2760
定　　价	280.00 元